TABLE OF CONTENTS

I. INTRODUCTION 1
   A. BACKGROUND 1
      1. Influences on Tropical 3
      2. Using the Large Scale
         on an Appropriate Scale 4
   B. SCOPE, LIMITATIONS, AND ASSUMPTIONS 6
      1. Scope of the Work 6
      2. Limitations 7
      3. Assumptions 13
         a. Assumptions 13
         b. Hypotheses 13
   C. OUTLINE 14

II. DATA AND METHODS 17
   A. STUDY PERIOD AND REGION 17
   B. DATA SOURCES AND SELECTION 17
   C. VARIABLES 24
      1. First Stage Independent Variables for Global
         Warming/Environmental Factors 24
      2. Response Variables for First Stage
         Independent variables (year) and
         Intermediate Stage Independent Variables 24
      3. Response Variables for Second Stage
         Independent Variables 25
   D. TEMPORAL-SPATIAL DATA BLOCKS 25
   E. STATISTICAL ANALYSES METHODS 27

III. ANALYSIS AND RESULTS 31
   A. LONG TERM TRENDS IN THE LARGE SCALE ENVIRONMENTAL
     FACTORS 31
      1. Sea Surface Temperature 31
      2. Shear 34
      3. Vorticity 37
      4. Mean Upward Vertical Velocity 39
      5. Relative Humidity 41
   B. LONG TERM TRENDS IN TROPICAL CYCLONE ACTIVITY 47
      1. Tropical Cyclone Frequency 47
      2. ACE 49
   C. MODELLING OF TROPICAL CYCLONE ACTIVITY IN THE
     WESTERN NORTH PACIFIC USING LARGE SCALE
     ENVIRONMENTAL FACTORS 54
      1. Tropical Cyclone Frequency 54
         a. The Effect of SST on TC Formation 62
         b. The Effect of Shear on TC Formation 64

    c. *The Effect of Relative Humidity on TC Formation* 65
    d. *The Effect of Vertical Velocity on TC Formation* 67
    e. *The Effect of Absolute Vorticity on TC Formation* 68
    f. *Model Validation* 69
    g. *Sensitivity of Model Probabilities to Each LSEF* 78
    h. *Relationship Between Global Warming and TC Formation* 79
  2. ACE 79
    a. *Developing a Predictive Model for ACE* ... 80
    b. *ACE Partial Residual Plots - SST* 84
    c. *Shear Partial Residual Plot for ACE* 85
    d. *Relative Humidity Partial Residual Plot for ACE* 86
    e. *Vertical Velocity Partial Residual Plot for ACE* 87
    f. *Absolute Vorticity Partial Residual Plot for ACE* 88
    g. *Sensitivity of Model ACE to the LSEFs* ... 89
    h. *Relationship Between Global Warming and ACE* 90

IV. CONCLUSIONS AND RECOMMENDATIONS 91
  A. TRENDS IN THE LARGE SCALE ENVIRONMENTAL FACTORS ... 91
  B. TRENDS IN TROPICAL CYCLONE ACTIVITY 92
  C. LESSONS LEARNED 96
  D. RECOMMENDATIONS FOR FUTURE STUDY 97

APPENDIX A. TC FORMATION PROBABILITY CONTOUR PLOTS FOR WEEKS 20-52 OF 2004 99

APPENDIX B. TC FORMATION PROBABILITY CONTOUR PLOTS FOR WEEKS 20-52 OF 1979 117

LIST OF REFERENCES 135

INITIAL DISTRIBUTION LIST 139

LIST OF FIGURES

Figure 1. Proposed regression modeling.....................6
Figure 2. Time series plot of CO_2 concentration (ppm) for the time frame 1970-2006, with twelve monthly data points per year............................8
Figure 3. Global Average Surface Temperature Anomaly (deg C multiplied by 100) vs. CO_2 Concentration (PPM)..9
Figure 4. GASTA vs. the author's age (years)..............12
Figure 5. Global averaged surface temperature anomaly (°C) with respect to 1901-2000 mean (from NOAA)..19
Figure 6. August 1997 SST contour plot (degrees C*100)....21
Figure 7. Contour plots of zonal wind speed at 200mb level based on data from the Reanalysis-II (top) and Reanalysis-I (bottom) data sets.......22
Figure 8. 1997 TC tracks in the WNP. Each dot marks the chronologically ordered position of a TC in six hourly intervals according to the best track data set. Note the number of recurring TCs (those that have a strong northward component). Because of recurring TCs, our WNP region extends to 40N...................................24
Figure 9. TC formation sites for 1970 - 2005, with our WNP region shown by the red box.................26
Figure 10. SSTs (°C*100) averaged weekly and over all five degree blocks inside the WNP vs. year. Note the increasing trend............................32
Figure 11. Box plots of SSTs over entire WNP using data for weeks 20-52.................................33
Figure 12. Conditional plot showing SSTs conditioned on latitude and longitude. Increasing trend in SST over time is evident, though different from region to region. The temperatures marked with green backgrounds are the total SST changes during 1970-2006 for the nine regions. The green background behind each regional SST change indicates that the change is considered favorable for increased TC activity...............34
Figure 13. Least squares fit of shear data using year only as a dependant variable. Note the very slight increase in average shear with time.............36
Figure 14. Shear vs. year conditioned on latitude and longitude. The change in shear (1970-2006) is

shown for each region with green background indicating the change is considered favorable for increased TC activity and red considered unfavorable..................................37

Figure 15. Absolute vorticity vs. year, conditioned on latitude and longitude. Note the predictor variable that dominates the response is latitude...39

Figure 16. Least squares fit of vertical velocity vs. year...40

Figure 17. Box plot of vertical velocity vs. year conditioned on latitude and longitude..........41

Figure 18. Least squares regression line based on both Reanalysis I and II data showing an increase in relative humidity over time......................42

Figure 19. Box plot of relative humidity showing discontinuity at the 1979-1980 transition between the Reanalysis I and II data sets.......43

Figure 20. Box plots of relative humidity conditioned on latitude and longitude. The discontinuity between data sets at 1980 is again readily apparent...44

Figure 21. Plot of least squares fit using Reanalysis II relative humidity data only......................45

Figure 22. Relative humidity box plots conditioned over latitude and longitude............................46

Figure 23. Least squares linear regression of TC formation for 1970-2005 in the Western North Pacific......48

Figure 24. Normal probability plot showing a normal distribution of the residuals...................49

Figure 25. ACE (kt^2) vs. year for the WNP, with linear trend included...................................51

Figure 26. Normal probability plot of the ACE residuals showing the residuals are indeed normally distributed..52

Figure 27. Average ACE per storm vs. year showing a small but pronounced upward trend......................54

Figure 28. Scatter matrix of time, main factors, CO_2 concentration, and temperature anomaly..........59

Figure 29. Contribution to TC formation probability model by the main effect shear........................60

Figure 30. Scatter matrix of relative humidity and vertical velocity................................61

Figure 31. Partial residual plot for SST as used in our logistic regression model.......................63

Figure 32. Partial residual plot for shear as used in our logistic regression model..........................65
Figure 33. Partial residual plot for relative humidity as used in our logistic regression model............66
Figure 34. Partial residual plot for vertical velocity as used in our logistic regression model............68
Figure 35. Partial residual plot for absolute vorticity as used in our logistic regression model............69
Figure 36. Model weekly probability contours for TC formation for week 43 of 1997. Red dots mark formation locations for actual TCs................71
Figure 37. Model weekly probability contours for TC formation for week 29 of 1997. Red dots mark formation locations for actual TCs................72
Figure 38. Model weekly probability contours for TC formation for week 50 of 1997. Red dots mark formation locations for actual TCs. Note the low probabilities compared to those in Figures 36-37. Note also that no TCs formed during this week..73
Figure 39. Model weekly probability contours for TC formation for week 43 of 1974. Red dots mark formation locations for actual TCs. TC data for 1970-1979 was not used in the model development. Thus, the results in this figure and in figures 39-42 indicate that there is a good fit between our model and actual TC formation in the Western North Pacific..........74
Figure 40. Model weekly probability contours for TC formation for week 42 of 1974. Red dots mark formation locations for actual TCs. TC data for 1970-1979 was not used in the model development. Thus, the results in this figure and in figures 39-42 indicate that there is a good fit between our model and actual TC formation in the Western North Pacific..........75
Figure 41. Model weekly probability contours for TC formation for week 39 of 1974. Red dots mark formation locations for actual TCs. TC data for 1970-1979 was not used in the model development. Thus, the results in this figure and in figures 39-42 indicate that there is a good fit between our model and actual TC formation in the Western North Pacific..........76
Figure 42. Model weekly probability contours for TC formation for week 20 of 1974. Red dots mark

	formation locations for actual TCs. TC data for 1970-1979 was not used in the model development. Thus, the results in this figure and in figures 39-42 indicate that there is a good fit between our model and actual TC formation in the Western North Pacific. Note the low probabilities compared to Figures 39-41 and the absence of any TC formations............77	
Figure 43.	Annual ACE values for 1970-2005. Actual (in blue) and modeled (in red)......................83	
Figure 44.	ACE partial residual plot for SST...............85	
Figure 45.	ACE partial residual plot for shear.............86	
Figure 46.	ACE partial residual plot for relative humidity..87	
Figure 47.	ACE partial residual plot for vertical velocity..88	
Figure 48.	ACE partial residual plot for absolute vorticity.......................................89	

LIST OF TABLES

Table 1.	A selection of potential LSEF data sources	18
Table 2.	A selection of prior studies and their cited data sources	18
Table 3.	Estimated coefficients and standard errors for model to predict the formation of a TC in a weekly five by five degree block	57
Table 4.	Relative contribution to formation probability given a 10% change in each of the LSEFs while holding the other LSEFs constant	79
Table 5.	Model coefficients and standard errors for ACE	81
Table 6.	Relative contribution to ACE given a 10% change in each of the LSEFs while holding the other LSEFs constant	90

THIS PAGE INTENTIONALLY LEFT BLANK

LIST OF ABBREVIATIONS AND ACRONYMS

AbsVort	Absolute vorticity (s^{-1})
ACE	Accumulated cyclone energy (kts^2)
ANOVA	Analysis of variance
ERSST	Extended reconstructed sea surface temperature
GASTA	Global average surface temperature anomaly (°C)
JMA	Japan Meteorological Agency
JTWC	Joint Typhoon Warning Center
latMax	Northernmost latitude of a weekly block
lonMax	Easternmost longitude of a weekly block
LSEF	Large scale environmental factor
km	Kilometers
kts	Knots
NCAR	National Center for Atmospheric Research
NCEP	National Centers for Environmental Prediction
nmi	Nautical mile
OISST	Optimum interpolation sea surface temperature
Pa/s	Pascals per second
ppm	Parts per million
RelHum	Relative humidity
ROC	Region of concern
s	Second
SST	Sea surface temperature (deg C)
TC	Tropical cyclone
VertVel	Vertical velocity (Pa/s)
WNP	Western North Pacific

THIS PAGE INTENTIONALLY LEFT BLANK

ACKNOWLEDGMENTS

First and foremost, I need to thank my wife Dionne, who has heard more about tropical cyclone activity over the past year than anyone should. Her love, support, patience, and encouragement were the foundation I needed in this endeavor. Thank you, Dionne.

This thesis would never have even been started without the support of the late Capt. Starr King, who was willing to let me take on a rather non-traditional topic for the OR Department, gave me a point of contact, and encouraged me to "tilt at windmills".

I am also indebted to the Operations Research Department Chairman, Dr. Jim Eagle, who strongly encouraged me to work on getting a Master of Science in Operations Research while a military faculty member in the Department. Without that push, I would have never been in the position to conduct this research.

There are many NPS faculty and staff who have earned my appreciation as well. The most notable of these are Bob Creasey who literally downloaded gigabytes of data and organized it into a format I could use, Mark Boothe who found an otherwise nearly impossible to find problem I had in my intensity data, Dr. Russ Elsberry, who made time for questions, introduced me to Dr. Greg Holland, and gave unvarnished feedback, Dr. Pat Harr, who also was very unselfish with his time and inundated me with links to many journal articles to start my literature review, and Dr. Dave Alderson, who gave great advice on selecting a programming language, as well as some "starter code" that enabled me to finally get going computationally.

I also must say thank you to my second reader, Dr. Sam Buttrey whose assistance with S-Plus was much appreciated.

Dr. Lyn Whitaker devoted much time, effort and advice to me as one of the co-advisors for this project. Her insight, ideas, technical wisdom, and "let's not panic" attitude weren't just helpful, they were invaluable to this project, and contributed to helping shape this project into what it is. Thank you.

Finally, it is not possible to thank enough Dr. Tom Murphree, my other co-advisor. He listened to some guy who couldn't even spell TC a year ago, who wanted to try using regression to show relationships between global warming and tropical cyclones. Thank you for listening, for the enormous amount of one on one teaching you have done, for your insight (we might have had very inconclusive results had we chosen any other scales), And For Your Editing, which is the only reason why this thesis is coherent.

EXECUTIVE SUMMARY

The purpose of this research is to identify the relationships between global warming and tropical cyclone (TC) activity in the Western North Pacific (WNP). The consensus among researchers who study TC activity is that TC activity (formation and strength) is a function of several large scale environmental factors (LSEFs). These factors are:

- Large values of low level absolute vorticity
- Weak vertical shear of the horizontal winds
- Sea surface temperatures (SSTs) exceeding 26 C
- Mean upward motion and
- High mid-level humidities

Our study focused on three main points. We wanted to verify that the LSEFs given above really do influence TC activity. We also wished to examine whether global warming is exerting a detectable influence on these LSEFs, and subsequently, whether a global warming induced change in these LSEFs is exerting a detectable influence on TC activity.

Data for these factors was gathered for the years 1970 (the beginning of the satellite era, ensuring our TC count is reasonably correct and consistent) through 2005 and averaged into a weekly grid of five by five degree blocks covering the WNP. Regression was performed to determine how these factors are changing with time. We determined that SSTs are increasing, as is shear, though in the tropics, where TC activity is greatest, the magnitude of shear is actually decreasing.

Logistic regression was used to examine the relationship between the LSEFs and the probability of TC formation. It was determined that all the previously mentioned LSEFs are important in influencing TC formation probability, with SST and vertical velocity being the factors to which TC formation is most sensitive. Using the logistic regression fit based on our 1980 - 2005 data, we estimated the individual TC formation probabilities by week for each of the blocks of our grid. We then displayed weekly estimated probability contour plots showing the probability of formation with actual TC formations superimposed. The performance of these probability contour plots against actual TC formations validates the model. The logistic regression model for the probability of TC formation is:

$$\log(\hat{p}/(1-\hat{p})) = -22.3677 + 0.0066 SST + 1.35 Coriolis + 0.0407 AbsVort + 0.1436 VertVel - 0.0276 RelHum - 0.0264 Shear - 0.0031 Shear^2$$

where $\hat{p}$ is the estimated probability of TC formation.

Likewise, linear regression and analysis of variance were used to determine what if any relationship exists between the LSEFs and accumulated cyclonic energy (ACE, a measure of the strength of a storm). Again, it was determined that all the previously mentioned LSEFs appear to influence ACE, with SST and absolute vorticity being the factors to which ACE is most sensitive. Our model was validated by comparing our predicted values of ACE to actual values; there was a surprisingly good agreement. Our validated model is:

$$ACE = -92365 + 26.52 SST + 3005.25 AbsVort + 58.83 VertVel + 211.99 RelHum + 55.32 Shear - 5.09 Shear^2$$

with a standard error of 20380 kts^2 on 4448 degrees of freedom.

Therefore it is our contention that through the LSEFs, global warming has and will continue to increase the average number and intensity of TCs in the WNP.

THIS PAGE INTENTIONALLY LEFT BLANK

I. INTRODUCTION

A. BACKGROUND

Global warming is a controversial topic. Is it for real? Is much being made of nothing? Is combating it worth the required significant societal and economic changes?

The possible manifestations of global warming are especially controversial. Particularly after hurricane Katrina hit the U.S. in August 2005, there has been much conjecture that recent changes in hurricane activity in the Atlantic[1] have been due to global warming[2]. The purpose of this study is to identify what (if any) environmental changes have been occurring that may have impacted tropical cyclone (TC) activity.

Changes in TC activity (we define TC activity as TC formation and TC intensity) have significant ramifications for U.S. military basing/staging; the health of the economy, and the U.S. population as a whole; and there are certainly ramifications for the entire "global village".

Prior studies show conflicting results. Klotzbach maintained there is no global warming – TC activity link[3]. Emanuel asserted that there is no discernable trend in TC formations per year, but that intensity is increasing[4]. Chan and Liu showed a relationship between TC intensity in the Western North Pacific and El Nino, but no relationship

---

[1] P. J. Webster, G. J. Holland, J. C. Curry, H.-R. Chang, *Science* **309**, 1844 (2005).

[2] R. A. Anthes, R. W. Corell, G. Holland, J. W. Hurrell, M. C. MacCracken, K. E. Trenberth, *Bull.Amer.Meteor.Soc* **87**, 623 (2006).

[3] P. J. Klotzbach, *Geophysical Research Letters*, **33**, L10805 (2006).

[4] K. Emanuel, *Nature*, **436**, 686 (2005).

between intensity and global warming[5]. Webster et al. concluded that TC formations and intensity are both increasing[6]. Pielke et al. make the point that the impact of TC activity on mankind is dominated by societal vulnerability, and not by increases in TC activity[7]. Certainly the focus of these studies has been on TC activity in the Atlantic. Wu et al. identified notable differences between several TC data sets for the WNP in terms of their depiction of changes in TC activity over the last several decades[8]. Landsea concluded that the available TC data is insufficient to allow any impacts of global warming on TC activity to be distinguished from variations in TC activity due to other causes[9] (e.g., natural multi-decadal climate variations). Trenberth and Shea asserted that global warming has resulted in increasing global average sea surface temperatures[10].

A common characteristic of these prior studies is that they focused on very large spatial and temporal scales. Emanuel (2005), for example, calculated a basin wide measure of TC intensity and compared it to tropical SST averaged over a very large portion of the same basin. Both quantities were smoothed twice before correlating them. Our concern is that this large scale averaging and

---

[5] J. C. L. Chan, K. S. Liu, *Journal of Climate*, **17**, 4590 (2004).

[6] P. J. Webster, G. J. Holland, J. A. Curry, H.-R. Chang, *Science,* **309**, 1844 (2005).

[7] R. A. Pielke Jr, C. Landsea, M. Mayfield, J. Laver, R. Pasch, *Bull.Amer.Meteor.Soc* **86**, 1571-1575 (2005)

[8] M.C. Wu, K.H. Yeung, W.L. Chang, *Eos,* **87**, 537-538 (2006).

[9] C.W. Landsea, *Eos,* **88**, 197-208 (2007).

[10] K. E. Trenberth, Dennis J. Shea, Geophysical Research Letters, **33**, (2006).

smoothing significantly dilutes the information necessary to understand changes in TC activity related to global warming.

1. Influences on Tropical Cyclone Activity

Over the past 50 years, significant strides have been made in the understanding of tropical cyclone activity. The consensus is that TC formations, intensities and tracks, are strongly influenced by several primary large scale environmental factors (LSEF), including[11]:

- Sea surface temperatures (SSTs) exceeding 26 C
- Weak vertical shear of the horizontal winds
- Large positive absolute vorticity at low levels
- Mean upward motion
- High mid-level humidity

The LSEFs are used to analyze and forecast TC activity, especially TC formations. Some seasonal predictions of TC activity in the North Atlantic are based on derived empirical relationships, including relationships between the LSEFs and TC formations[12].

In the context of trying to show a relationship between TC activity and global warming, the state of the art is to use these factors on a very grand scale. In some cases, SSTs, by far the most tangible of the LSEFs, are used almost exclusively for analyzing and predicting TC activity, with the remaining LSEFs being virtually ignored.

---

[11] W. M. Frank, 1987, Tropical Cyclone Formation. Chap. 3, *A Global View of Tropical Cyclones.* Office of Naval Research, Arlington, VA 22217, 53-90.

[12] W. M. Gray, 1975: *Tropical Cyclone Genesis in the Western North Pacific*, Naval Air Systems Command, Washington DC, 20361.

2. Using the Large Scale Environmental Factors on an Appropriate Scale

We think that much is lost in the understanding of TC activity if, as in prior studies, only one of the LSEFs (e.g., SST) is used to assess the impacts of global warming on TC activity, and if an LSEF is analyzed using data averaged over millions of square kilometers and several months, seasons, or years. The prior studies based on these sorts of analyses have had mixed results[13,14]. Contrary to the approach of these prior studies, we contend that all of the LSEFs are important in determining the impacts of global warming on TC activity, and that they do so on relatively local spatial and temporal scales (with respect to TC scales). In other words we think a tropical cyclone forms because the LSEFs on scales of several hundred kilometers, and over a few days or up to a couple of weeks are favorable for TC formation, not because the average temperature for the tropical North Atlantic for the year 2002, smoothed out over five years exceeded 26° C.

Therefore, we decided to study the relationship between global warming and each of the LSEFs to determine if any predictions regarding TC activity can be made given the effect of global warming on each LSEF, if indeed there appears that there is such an effect. We also decided to analyze the LSEF and TC data at space and time scales consistent with the scales at which the LSEFs are thought to influence TC activity (i.e., scales of about a week and hundreds of kilometers, rather than scales of a whole TC season (about six months) and a whole tropical basin (thousands of kilometers)). Thus in this study, we test

---

[13] K. Emmanuel, *Nature*, **436**, 686-688 2005).

[14] Johnny C. L. Chan, *Science*, **311** 2006).

the hypothesis that if global warming is influencing TC activity, it is doing so through all the LSEFs, and it is doing so at scales of hundreds of kilometers and about a week.

Most of the data on the LSEFs is available in six hour time intervals and in two and a half by two and a half degree latitude by longitude geographical increments (about 250 by 250 km squares). This allowed us to analyze the LSEFs on scales that fit their known impacts to TC activity. This should allow us to confirm that the LSEFs have an impact on TC activity, if they do.

The main purpose of the regression models is to facilitate the identification of relationships between global warming and TC activity, through the LSEFs commonly understood to play a role in TC activity (Figure 1). However, it is possible that these models might also be useful in analyzing and forecasting TC activity in general.

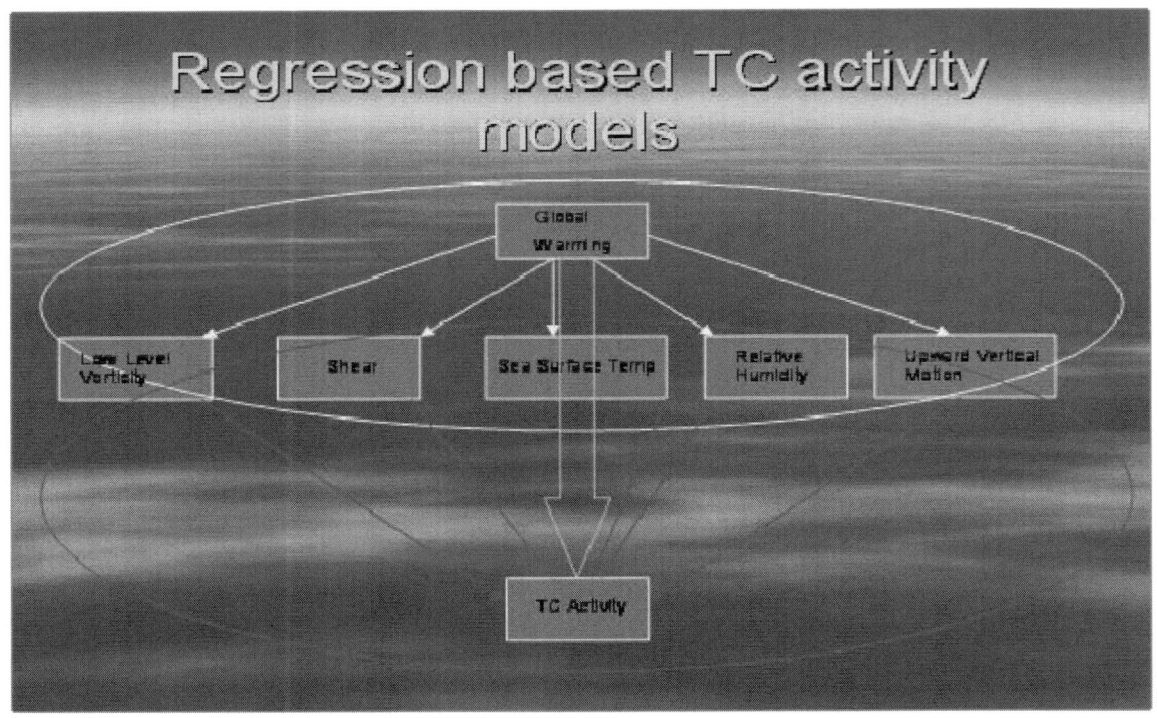

Figure 1. Proposed regression modeling.

B. SCOPE, LIMITATIONS, AND ASSUMPTIONS
1. Scope of the Work

This thesis will build on the prior studies of the climatology of TC activity, and the relatively limited work done investigating TC-global warming relationships.

The research conducted will focus on the Western North Pacific (WNP) for many reasons. The WNP is:

1. the region with the largest annual number of TCs (approximately 35 percent of the global total)
2. a region where TC activity has demonstrated effects (via teleconnections) on the U.S., and even Europe
3. bordered by nations which comprise a large portion of the Earth's population
4. a region of interest to the U.S. military

Data used will include all available data during 1970-2006, from the onset of extensive satellite coverage to the present. Data prior to 1970 will not be used, since prior to full satellite coverage, there were many likely TCs that never made landfall and for which little or no data was collected. Before the satellite era, TCs at sea were documented as they were observed by ship or aircraft, or if they made landfall. Between 1945 and 1970 an increase in reported TC formations is quite possibly a result of an increasing number of ships and aircraft at sea. As there have been many contentious arguments over the TC data base, we deliberately chose to use data from the satellite era only to avoid this issue as best as we could[15].

2. Limitations

There are two choices for measures of global warming to be used as independent variables, atmospheric CO_2 concentration in parts per million (ppm) (see Figure 2), or globally averaged surface temperature anomaly (GASTA). CO_2 concentration is a rather direct cause of global warming, but it does not capture the whole story of greenhouse emissions. GASTA of course is not a cause of global warming, but rather an attempt to quantify it. The limitation of GASTA is that while there is little debate as to whether Earth is experiencing a warming period, there is much more debate as to whether that warming is due to anthropogenic (man-made) causes. Figure 3 shows the existing relationship between CO_2 concentration and GASTA from 1970 through 2006.

---

[15] C.W. Landsea, *Eos,* **88**, 197-208 (2007).

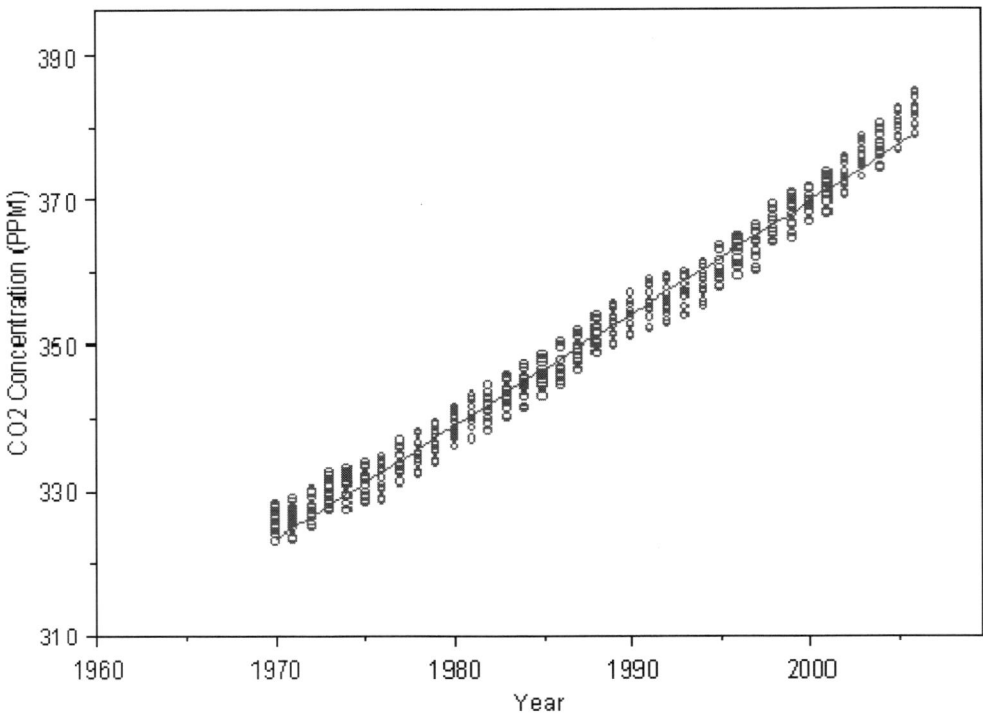

Figure 2. Time series plot of CO_2 concentration[16] (ppm) for the time frame 1970-2006, with twelve monthly data points per year.

The basic science behind greenhouse gasses and global warming is well established and understood. By altering the make up of the Earth's atmosphere by adding combustion gasses (primarily CO_2) from industry, power plants, and vehicles, the radiation of energy from Earth to space is changed in a way that causes Earth to get warmer. What is far less understood is what happens to Earth as it gets warmer. Earth is an enormous mechanical system, with a myriad of feedback cycles that operate on scales of minutes to millennia, making it very difficult to predict the

---

[16] NOAA website: http://www.esrl.noaa.gov/gmd/ccgg/trends/.

results of adding greenhouse gasses to the atmosphere (and also making simplistic predictions of the results of global warming laughably wrong).

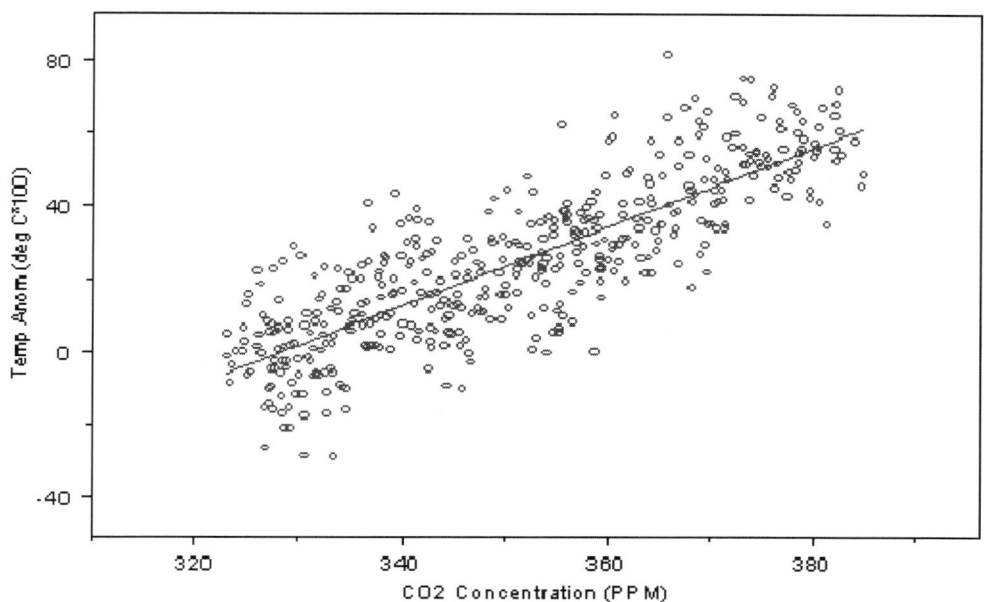

Figure 3. Global Average Surface Temperature Anomaly[17] (deg C multiplied by 100) vs. CO_2 Concentration (PPM).

Figures 2 and 3 show that CO_2 concentration is increasing approximately linearly with time, as is GASTA, and that there is a linear relationship between CO_2 concentration and GASTA. In our study, we have not attempted to determine the causes of global warming. Our work is focused on the relationships between observed global warming and observed TC activity in the WNP, not on the causes of global warming, natural or otherwise. One other clarification: as we have shown a linear relationship between CO_2 concentration and year, and a linear

---

[17] *Global Average Surface Temperature Anomalies*, http://lwf.ncdc.noaa.gov/oa/climate/research/anomalies/anomalies.html.

relationship between CO_2 concentration and GASTA, we will use years as a proxy measure of global warming. We will use years explicitly as an independent variable to show its relationship with each of the LSEFs and with WNP region wide TC activity. The purpose of using years as a proxy is to aid the reader, since GASTA is a less tangible independent variable than years.

Our study will not establish <u>causal</u> relationships. However, we do expect the evidence that we develop in this study will contribute to a determination of the causes of any long term changes in the LSEFs and in TC activity in the WNP. The determination of these causes will require the application of many different methods, including statistical analyses of observational data (such as ours), theoretical studies, and modeling studies.

Our study will focus on identifying the relationships between the LSEFs and (a) our proxy measure of global warming and (b) TC activity. For our analysis of the effects of the LSEFs on TC activity, the LSEFs will be our direct independent variables and TC formation and accumulated cyclone energy (ACE) will be our response variables. ACE quantifies TC intensity by squaring TC wind velocity and summing at six hourly intervals. For our analysis of the impacts of global warming on TC activity, the LSEFs will serve as our intermediate response variables, with our final response variables being TC frequency and ACE. It is important to keep in mind that the statistical relationships we will show do not by themselves allow us to determine causality. However, the relationships may contribute to the determination of causality, as discussed in the preceding paragraph.

An example is in order. A regression analysis between atmospheric CO_2 concentration (which has been steadily increasing during 1970-2006) and GASTA over the time frame (using the statistical software package *S plus*) gives:

Temp(°C*100) = -358.25 + 1.091CO_2Concentration   (1)

with a standard error of 12.28 on 1922 degrees of freedom.

This result estimates that for every 1 ppm increase in atmospheric CO_2, mean GASTA increases by 0.011 degrees C, with a standard error of 0.017. The adjusted R^2 is 0.69 which measures the strength between GASTA and CO_2 concentration.

The result is a surprisingly strong relationship that might suggest irrefutable proof that greenhouse gasses are causing an undeniable warming of the planet; certainly the strength of the relationship ought to be part of the global warming discussion. Figure 4 however, shows the relationship between the author's age (which unfortunately also has been steadily increasing with time) and GASTA. Regression analysis exploring the relationship between the author's age and GASTA yields the following equation:

Temp(°C*100) = -15.5 + 1.68Age   (2)

with a standard error of 12.6 on 1922 degrees of freedom.

Now the results estimate that for every year increase in my age, mean GASTA increases by 0.017 degrees C, with a standard error of 0.026 and an adjusted R^2 of 0.67. While the relationship is strong, it is unlikely that my age is causing global warming.

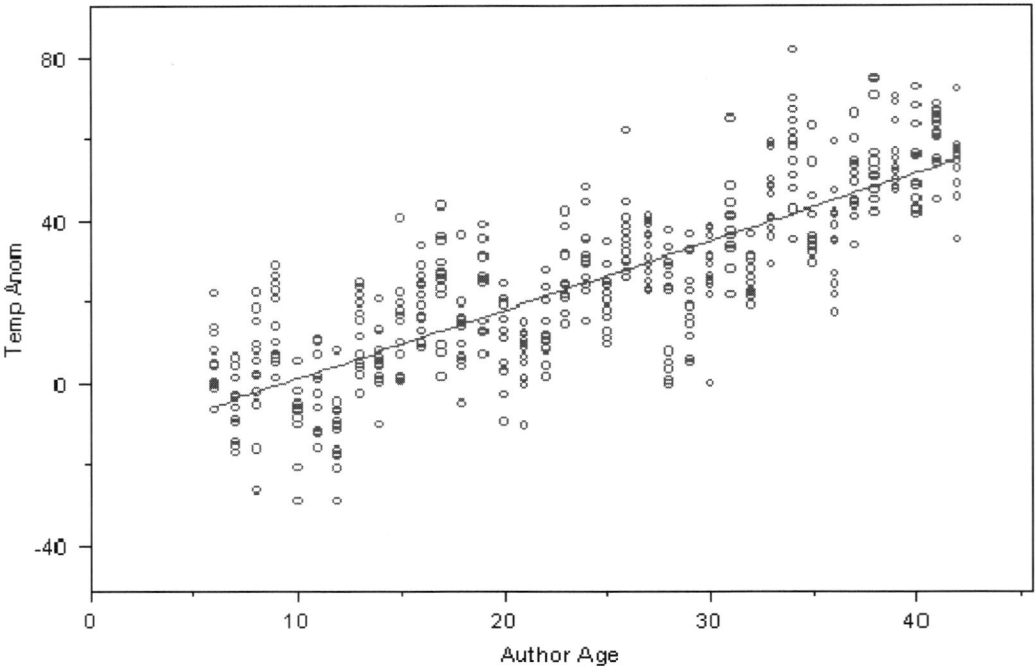

Figure 4. GASTA vs. the author's age (years).

In other words, regression can show the strength of a relationship between predictor and response variable, but additional information is needed to establish whether the relationship is physically plausible and therefore indicative of causality. In this case, reasoning based on additional information from atmospheric science theory and modeling indicates that global warming over several decades *may* influence TC activity. Therefore, a statistical relationship between global warming and TC activity may be physically plausible and indicative of causality; thus the relationship is relevant to the determination of causality. However, there are no theoretical or modeling indications that the author's age is related to changes in TC activity.

We will spare the reader the regression we did showing an inverse relationship between hair follicle density on his head and global warming.

3. Assumptions

Initial assumptions and hypotheses made in support of this work include the following:

a. *Assumptions*

1. That the spatial and temporal resolution of the data is sufficient to answer our questions.
2. Our global warming index, years, sufficiently captures the major signals of global warming, and impacts (or does not impact) the LSEFs, which in turn influence TC activity, in such a way as to be useful when using regression tools.

Our work will extend the prior research with the goal of answering the following questions with analysis based on local values of the LSEFs and TC activity rather than values averaged over large regions of both time and space as has been done in previous work.

1. In the WNP, does there appear to be a relationship between global warming and any of the LSEFs which influence TC activity?
2. In the WNP, does there appear to be a relationship between global warming and TC activity? If so, for what aspects does global warming appear to have impacts?
3. Can a relationship between the LSEFs be modeled and used to establish relationships between global warming and TC activity?
4. If relationships between global warming and the LSEFs can be identified, and if the relationships between the LSEFs and TC activity can be adequately modeled, then can the impacts of global warming on TC activity be modeled and potentially predicted?

b. *Hypotheses*

The above questions will be addressed in the process of exploring the following hypotheses:

1. Any relationship between global warming and TC activity is because global warming is influencing the LSEFs, and because the LSEFs influence TC activity.
2. TC activity is a function of local values of the LSEFs not values averaged over large regions of time or space.
3. TC formations are a function of each of the LSEFs.
4. ACE is a function of the each of LSEFs.

Hypothesis testing will be performed as follows:

1. Using least squares we will show the nature of the relationship between each of the LSEFs (each being a dependent variable) and global warming (where year is our proxy and the independent variable). This will support the hypothesis that global warming influences the LSEFs.
2. Using the tools of regression, a TC formation probability model will be developed using local values of each of the LSEFs. The model will be validated by comparing performance of the model to actual TC formation locations.
3. Using the tools of regression, an ACE model which estimates ACE given a TC has formed will be developed using local values of each of the LSEFs. The model will be validated by comparing ACE totals estimated by the model to actual ACE totals for each year.

C. OUTLINE

Chapter II briefly discusses the data sources available and the data sets used to complete this work, as well as the rationale as to why particular data sources were chosen. The methods for error checking the raw data and converting the data into a data set designed for ease of using regression techniques are also discussed. Sample plots to demonstrate error checking are included. Finally, we define the WNP region for the purposes of our work, and explain the rationale for choosing this area.

Chapter III begins with addressing what large scale region wide changes can be seen in the LSEFs. These changes are then revisited on a smaller geographical scale to determine if the observed changes are uniform or have some sort of regional dependence. Large scale region wide trends in TC activity, both frequency of formation and ACE, also are addressed. Following the large scale region wide analyses, the results from regression models for TC formation probability and ACE are discussed, along with model validation results. Simple sensitivity analysis results for TC formations and ACE are shown as well.

In Chapter IV we outline the many conclusions and recommendations we developed from the model results presented in Chapter III. Areas for further research are discussed as well.

THIS PAGE INTENTIONALLY LEFT BLANK

II. DATA AND METHODS

A. STUDY PERIOD AND REGION

The time period we selected for our analysis is 1970 through 2006. This time period was chosen to maximize the size of the data set while utilizing data only from the WNP satellite era. While TCs form year round in the WNP, formations peak in the months of July through October and are at their minimum in February[18] so to control the size of our data set we only used data from weeks 20 through 52 (mid-May through December).

The region selected for our study was 0-40 North and 105-185 East. This region was chosen because it is consistent with areas defined as WNP by other researchers,[19] it minimizes the area that could include Eastern North Pacific TCs, and it allows for the common northward recurvature of many WNP TCs.

B. DATA SOURCES AND SELECTION

There is a large amount of data available to quantify both the LSEFs, and TC activity. Table 1 shows a listing of data sources, with the data type, periodicity, temporal coverage, and spatial coverage. Table 2 shows data used in a selection of prior studies of the links between global warming and TC activity.

---

[18] S. J. Camargo, A. W. Robertson, S. J. Gaffney, P. Smyth, www.datalab.uci.edu/papers/camargo_etal.pdf

[19] J. C. L. Chan, K. S. Liu, *Journal of Climate*, **17**, 4590 (2004).

Data set	Coverage (years)	Grid	Resolution	SST	Velocity	Vorticity	Humidity	Upward Velocity
NCEP I	1948-2006	Lat/lon	2.5X2.5 deg, 6 hourly		X	X	X	X
NCEP II	1979-2006	Lat/lon	2.5X2.5 deg, 6 hourly		X	X	X	X
(NCEP) OISST	1982-2006	Lat/lon	1X1 deg, 6 hourly	X				
(NCEP) ERSST	1854-2006	Lat/lon	2x2 deg, monthly	X				
Hadley Centre	1854-2006	Lat/lon	1x1 deg, monthly	X				

Table 1.   A selection of potential LSEF data sources.

Prior studies	LSEF data	TC activity data
Chan and Liu (2004)	NCEP	JTWC
Webster et al. (2005)	Not cited	Not cited
Emanuel (2005)	Hadley Centre	JTWC
Trenberth and Shea (2006)	Hadley Centre	N/A (Atlantic focus)
Klotzbach (2006)	NCEP	JTWC
Mann and Emanuel (2006)	Hadley Centre	N/A (Atlantic focus)

Table 2.   A selection of prior studies and their cited data sources.

Part of our approach in selecting the data for our study was to use, as much as possible, data that had been used in the key prior studies of global warming links to TC activity, in order to allow relatively direct comparisons of our methods and results to these prior studies.

We chose two measures for global warming, CO_2 concentration (see Figure 2) obtained from the National Oceanic and Atmospheric Administration[20], and global average

---

[20]*Trends in Atmospheric Carbon Dioxide - Mauna Loa,* http://www.cmdl.noaa.gov/ccgg/trends/.

surface temperature anomaly (GASTA), obtained from the National Climatic Data Center[21]. Figure 5 shows a time history of GASTA.

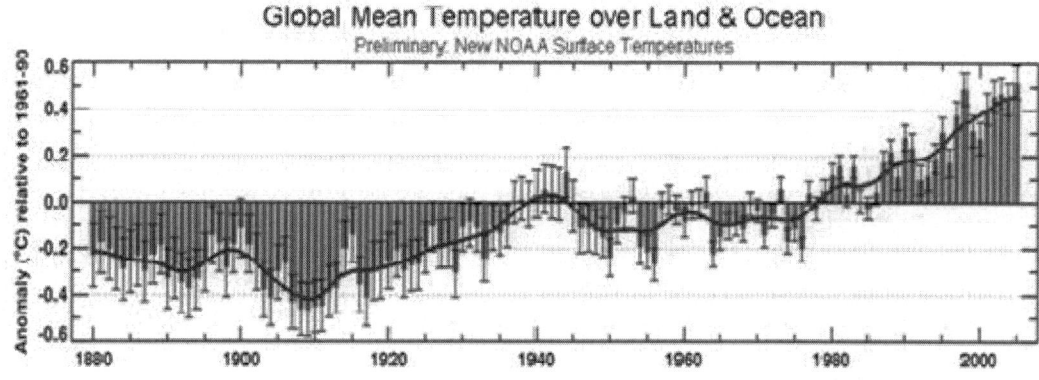

Figure 5. Global averaged surface temperature anomaly (°C) with respect to 1901-2000 mean (from NOAA).

CO_2 concentration as measured at Mauna Loa, Hawaii, is available at monthly intervals and certainly is a potential, although simplistic, predictor for global warming. GASTA is also available in monthly increments, and while it is a direct measure of the warming of the Earth, the extent to which it represents anthropogenic forcing is unclear[22]. GASTA may also cause complications in that when used as an independent variable, it may not be sufficiently independent of one of our response variables, SST. However, based on the relationships shown in Chapter I, years will be used as a proxy measure of global as we are more interested in changes in TC activity per year than changes in TC activity per ppm CO_2.

---

[21]*Global Surface Temperature Anomalies,* http://www.ncdc.noaa.gov/oa/climate/research/anomalies/anomalies.html.

[22]*Intergovernmental Panel on Climate Change, Working Group I,* http://ipcc-wg1.ucar.edu/wg1/wg1-report.html.

SST data are from the Hadley Center "ISST1" data set[23], from the UK Met Office website at http://hadobs.metoffice.com/hadisst/. This data is at a spatial resolution of one degree by one degree, and a monthly temporal resolution. The National Centers for Atmospheric Prediction (NCEP)/National Center for Atmospheric Research (NCAR) reanalysis SST data set was not selected because we wanted to be consistent with prior studies, the Hadley Centre data was at a finer spatial resolution, and because the NCEP OISST data set, while at a one degree and six hourly resolution, was not available for the full study period. The Hadley Centre ISST1 data set has been used by other researchers (Table 2), and is at the same temporal and better spatial resolution as the NCEP ERSST data so we chose to use the ISST1 set. Figure 6 is a sample contour plot of data.

---

[23] N. A. Rayner, D. E. Parker, E. B. Horton, C. K. Folland, L. V. Alexander, D. P. Rowell, E. C. Kent, A. Kaplan, Global analyses of sea surface temperature, sea ice, and night marine air temperature since the late nineteenth century J. Geophys. Res.Vol. 108, No. D14, 4407 10.1029/2002JD002670.

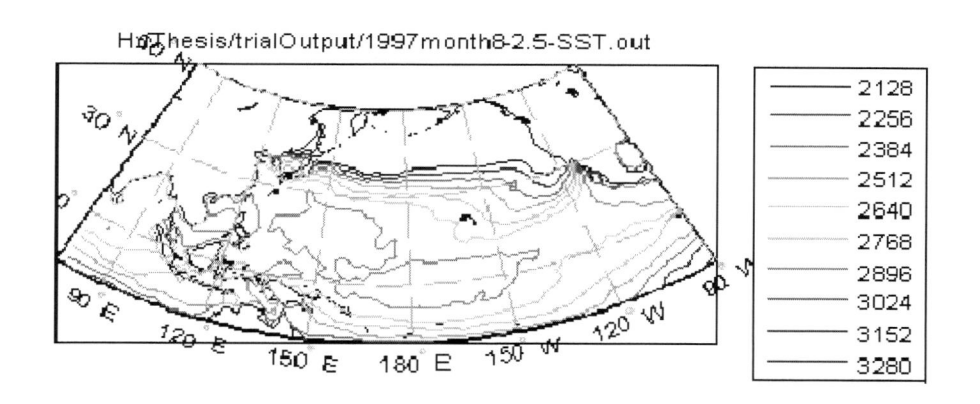

Figure 6. August 1997 SST contour plot (degrees C*100).

For the atmospheric LSEFs in the WNP, we chose the NCEP/NCAR reanalysis fields. Two sets of reanalysis fields were used: Reanalysis I, which is populated by variables taken at 2.5 by 2.5 degree increments on six hour intervals starting from 1948 and continuing to present,[24] and the Reanalysis II fields which start in 1979 and continue to present using the same temporal and spatial coverage as the Reanalysis I fields. Because the Reanalysis II fields are an improvement over Reanalysis I[25], the Reanalysis II fields were used over the entire timeframe for which they are available. The Reanalysis I field data were only used for 1979 and earlier. Since there is overlap between the two data sets, we compared

---
[24] Kalnay et al.,The NCEP/NCAR 40-year reanalysis project, *Bull. Amer. Meteor. Soc.*, 77, 437-470, 1996.

[25] *NCEP-DOE Reanalysis II: Summary,* http://www.cdc.noaa.gov/cdc/data.ncep.reanalysis2.html#detail.

data from both fields (e.g., Figure 7) and decided that little if any significant error would be introduced by using the two different data sources for most aspects of our study. The NCEP Reanalysis data was provided by the NOAA/OAR/ESRL PSD, Boulder, Colorado, USA, from its web site at http://www.cdc.noaa.gov/.

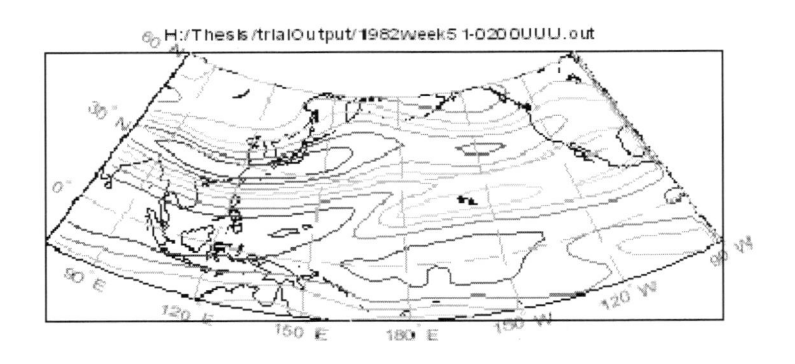

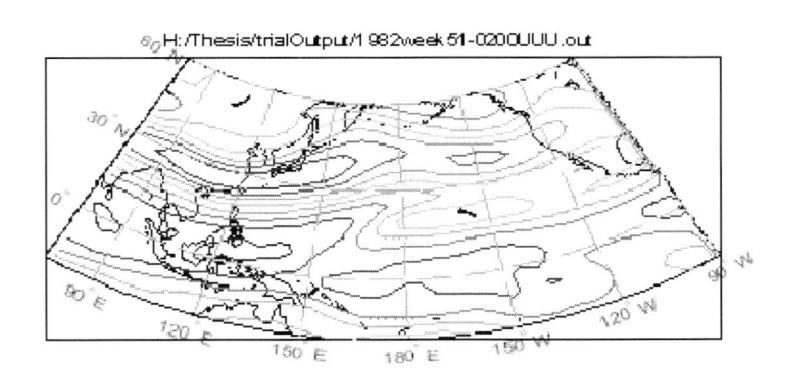

Figure 7. Contour plots of zonal wind speed at 200mb level based on data from the Reanalysis-II (top) and Reanalysis-I (bottom) data sets.

For our measures of TC activity, we chose to use tropical cyclone best track data (position, date and time, and wind velocity)[26] from the Joint Typhoon Warning Center (JTWC). The TCs included are all "wp" labeled best tracks from the JTWC, not just those that originate inside of our WNP region. There are several sources of TC data for the WNP, and some potentially important differences between these sources[27]. This has led to disagreement about what has actually happened with TC activity over the last several decades[28]. Our selection of the JTWC best track data set was based on it having been used in the preponderance of the literature reviewed. Figure 8 shows the JTWC analysis of the WNP TC tracks for 1997.

---

[26] *Joint Typhoon Warning Center Products* https://metocph.nmci.navy.mil/jtwc.php.

[27] M.C. Wu, K.H. Yeung, W.L. Chang, *Eos,* **87**, 537-538 (2006).

[28] C. W. Landsea, Eos, **88**, 197-208 (2007).

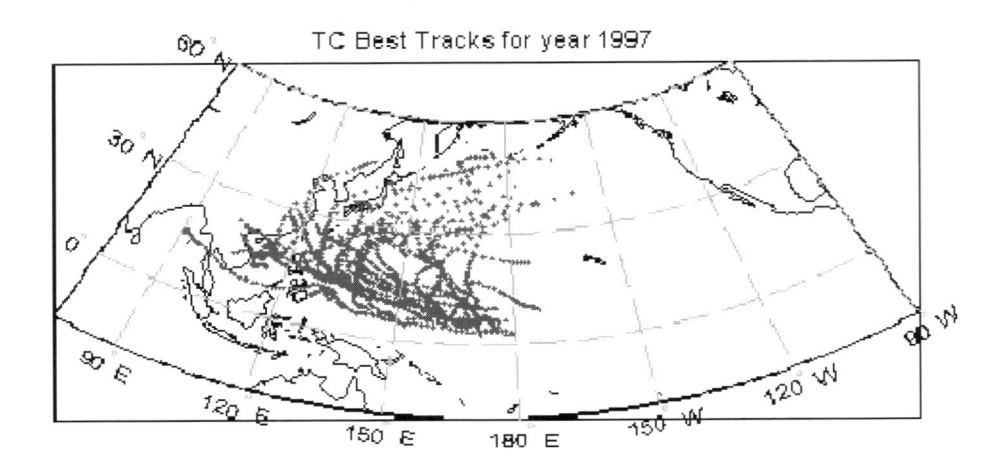

Figure 8. 1997 TC tracks in the WNP. Each dot marks the chronologically ordered position of a TC in six hourly intervals according to the best track data set. Note the number of recurving TCs (those that have a strong northward component). Because of recurving TCs, our WNP region extends to 40N.

C. **VARIABLES**

   1. **First Stage Independent Variables for Global Warming/Environmental Factors**

      - CO_2 concentration (ppm)
      - Global averaged surface temperature (°C)
      - Year (used as a proxy for the variables above based on strong linear relationships)

   2. **Response Variables for First Stage Independent variables (year) and Intermediate Stage Independent Variables**

      - Vorticity ($s^{-1}*10^5$) measured at 850 mb
      - Vertical wind shear (m/s), the 200 mb wind velocity minus the 850 mb wind velocity
      - Sea Surface Temperature (°C*100)

- Relative Humidity (%) at 500 mb
- Vertical Velocity (-1*Pa/s) at 500mb Note that as air ascends, it experiences lower and lower atmospheric pressure, thus pressure (in Pascals) decreases as ascent occurs. We multiplied the NCEP reanalysis vertical velocities by -1. Thus, in our results, positive vertical velocities correspond to upward air motions.

3. **Response Variables for Second Stage Independent Variables**

- TC formation (number)
- Accumulated cyclone energy (ACE, kts^2) a measure of the maximum sustained TC winds[29]. Only data for TCs that reached tropical storm intensity (maximum sustained winds of 18 m/s (35 kts) or greater) were used in our calculations of ACE.

D. **TEMPORAL-SPATIAL DATA BLOCKS**

Data was obtained for the period 1970-2006 (however, as of the end of this study, the 2006 JTWC best track data was not available). We averaged the six hourly NCEP reanalysis data into one week intervals. The purpose of this averaging was to bin the data into periods approximately consistent with those associated with the effects of the LSEFs on TC activity[30], keep the data set to a manageable size, and keep the program run time (for manipulating the data set) to a practical amount. We felt that operating on a weekly time scale would not unduly dampen the LSEF effects critical for TC activity and would be a vast improvement in temporal resolution over that in prior studies[31,32].

---

[29] Philip J. Klotzbach, Geophysical Research Letters, **33**, (2006).

[30] W. M. Gray, 1975: *Tropical Cyclone Genesis in the Western North Pacific*, Naval Air Systems Command, Washington DC, 20361.

[31] M. A. Mann, K. A. Emanuel, *Eos,* **87**, *233 (2006)*.

[32] J. C. L. Chan, K. S. Liu, *Journal of Climate*, **17**, 4590 (2004).

We also averaged the data into five degree latitude by five degree longitude blocks (about 500km by 500km) to represent the spatial scale at which the LSEFs are thought to influence TC activity[33]. Thus, the data was analyzed in temporally and spatially averaged blocks covering one week and five by five degrees per block. The end result is that over our defined area of the WNP (see Figure 9) we have 145,000 data blocks that include over 1000 TC formations and 4500 blocks where there was TC activity i.e., either a TC formation and/or positive ACE.

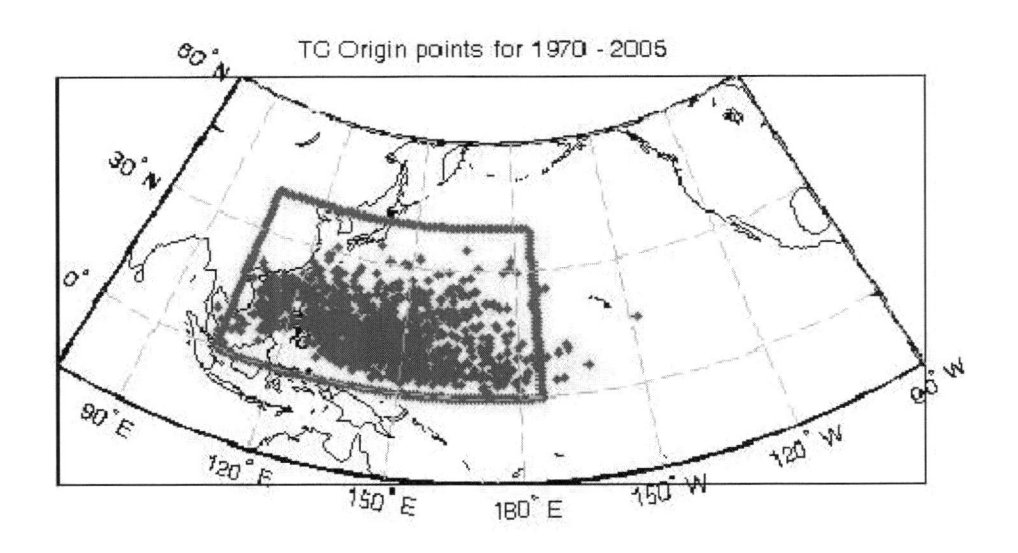

Figure 9. TC formation sites for 1970 - 2005, with our WNP region shown by the red box.

---

[33] W. M. Gray, 1975: *Tropical Cyclone Genesis in the Western North Pacific*, Naval Air Systems Command, Washington DC, 20361.

E. STATISTICAL ANALYSES METHODS

Several statistical methods were used in analyzing the data identified above. Least squares fitted lines[34] were used to identify any trends on a broad scale across the WNP. This was our tool chosen to evaluate the relationship between global warming and the LSEFs, and to estimate LSEF changes using the independent variable, years. Likewise, least squares were used on a broad scale to estimate the change in TC activity over our time period, again using years as our independent variable.

In addition, trellis plots were used to re-examine the behavior of the LSEFs. The trellis plots are used to explore the same data used for the least squares fits, except that the data was partitioned into smaller geographical sub regions as opposed to being analyzed for the overall WNP region. By use of trellis plots, we were able to examine whether any changes to the LSEFs were occurring uniformly across the WNP or whether there was, as we suspected, some sort of regional dependence on how the LSEFs changed. Seeing a regional dependence would provide some indication that analyzing our data on the small, weekly five by five degree scale might be more sensitive than using larger areas and periods, as done in prior studies of the effects of global warming on TC activity (see Chapter I).

Logistic regression[35] was used to estimate the relationship between the log-odds of TC formation, and the

---

[34] J. L. Devore, 2003: *Probability and Statistics for Engineering and the Sciences* 6th ed., Brooks/Cole-Thomson Learning, 10 David Dr., Belmont, CA 94002.

[35] D. Collett, 1991, Modelling Binary Data. Chap. 5, *Overdispersion*. Chapman and Hall, 2-6 Boundary Row, London SE1 8HN, UK, 188.

LSEFs. Our blocks of LSEF data were used to estimate an equation whereby the log-odds of TC formation probability is a linear function of the LSEFs. Using independent data, in weekly five by five degree blocks, we could then check and validate the performance of this model against expected behavior (e.g., the monsoon trough, a region in the tropical WNP where the LSEFs are often favorable for TC formation[36]) and against actual TC formations.

Multiple linear regression[37] was used to estimate the relationship between expected ACE and each of the LSEFs. Our blocks of LSEF data were used to estimate an equation whereby the expected amount of ACE generated in a weekly five by five degree block given a TC has formed is a linear function of the LSEFs. Using independent data in weekly five by five degree blocks, we could then check and validate the performance of the model against actual ACE generated in a given year. These analyses result in the following chain:

1. We used least squares to identify any broad scale relationships between global warming and the LSEFs. This analysis helped us to test the idea that global warming, if it has impacted TC activity, has done so via the LSEFs.
2. We examined whether broad scale changes in the LSEFs had any regional dependence. This helped us test the idea that the impacts of global warming need to be analyzed at scales comparable to those at which the LSEFs affect TC activity.
3. We used logistic regression applied to our small scale weekly five degree by five degree blocks to

---

[36] R. L. Elsberry, W. M. Frank, G. J. Holland, J. D. Jarrell, R. L. Southern, 1971: "A Global View of Tropical Cyclones", Office of Naval Research, Arlington, VA 22217.

[37] J. L. Devore, 2003: *Probability and Statistics for Engineering and the Sciences 6th ed.*, Brooks/Cole-Thomson Learning, 10 David Dr., Belmont, CA 94002.

estimate the relationship between the LSEFs and TC formation. This analysis helped us to test the idea that global warming, if it has impacted TC formations, has done so via the LSEFs.
4. We used multiple linear regression applied to our small scale weekly five degree by five degree blocks to estimate the relationship between the LSEFs and ACE. This analysis helped us to test the idea that global warming, if it has impacted TC intensity, has done so via the LSEFs.

THIS PAGE INTENTIONALLY LEFT BLANK

III. ANALYSIS AND RESULTS

A. LONG TERM TRENDS IN THE LARGE SCALE ENVIRONMENTAL FACTORS

1. Sea Surface Temperature

In concert with the trend observed in Figure 5 and as discussed by Trenberth and Shea[38], SSTs in the WNP have steadily increased with time during 1970-2006 (see Figure 10). The trend line in the graph is a least squares fit, with the equation for SSTs being:

$$\text{SST } (°C*100) = -338.5859 + 1.5138*\text{year} \quad (3)$$

yielding an estimated 0.529 degree centigrade increase in mean SST during 1970-2006. The standard error for the intercept and the year coefficients are 184.80 and 0.093 respectively. The coefficient for our year term, +1.5138 is an estimate of the relationship between year and SST, and has units of °C*100/year. As year is our proxy for global warming, equation (3) indicates that there is a positive relationship between global warming and WNP SSTs (but, as discussed in Chapter I, this relationship does not indicate causality, although it may contribute to a determination of causality). Note too, that the coefficient scales the equation (our SSTs are in units of °C*100) and keeps the units consistent, so the size of the coefficient is not indicative of the importance of the term that it scales.

---

[38] K. E. Trenberth, Dennis J. Shea, Geophysical Research Letters, **33**, (2006).

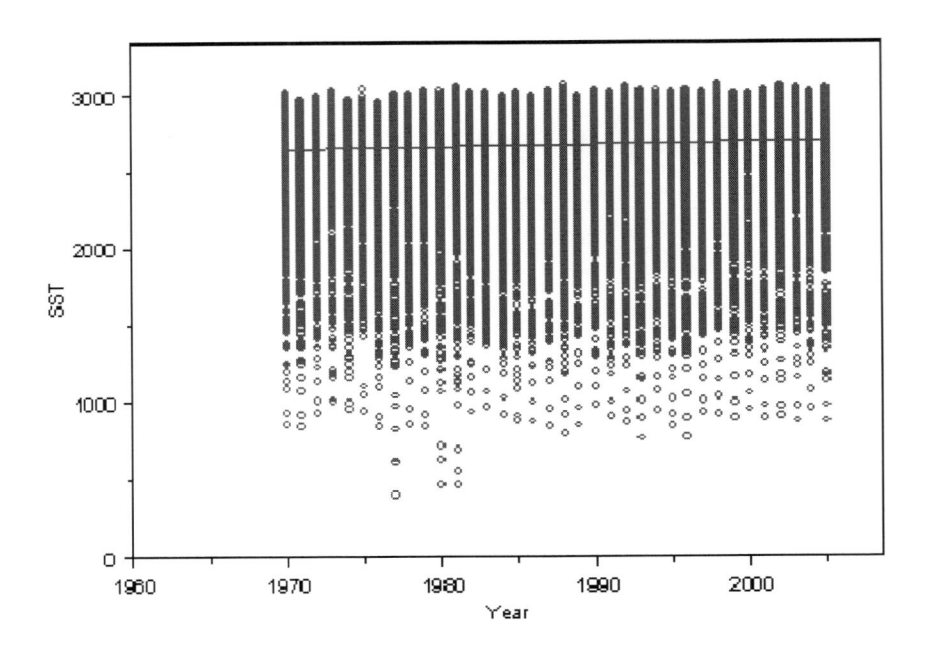

Figure 10. SSTs (°C*100) averaged weekly and over all five degree blocks inside the WNP vs. year. Note the increasing trend.

Realize that Equation 3 quantifies SST over the entire WNP. While Figure 10 aids in an overall understanding of what is happening with the SST LSEF, we remind the reader that we are more interested in what is happening in the smaller scale of weekly time frames and five degree blocks, since we expect that the LSEFs on these smaller scales will tell us more about TC activity.

Figure 11 below shows box plots of the same overall WNP SST data used above. However, Figure 12 shows that the behavior of SST clearly is dependant on both latitude and longitude (week is still being ignored here for the sake of simplicity). Positive trends are still apparent in all nine regions of the WNP shown in Figure 12. But the behavior of the SSTs even in these large regions varies considerably from region to region. This indicates that characterizing the behavior of SST and other LSEFs using WNP wide averages (averages over millions of square km) can mischaracterize what is happening in the vicinity of TC formation sites or TC tracks.

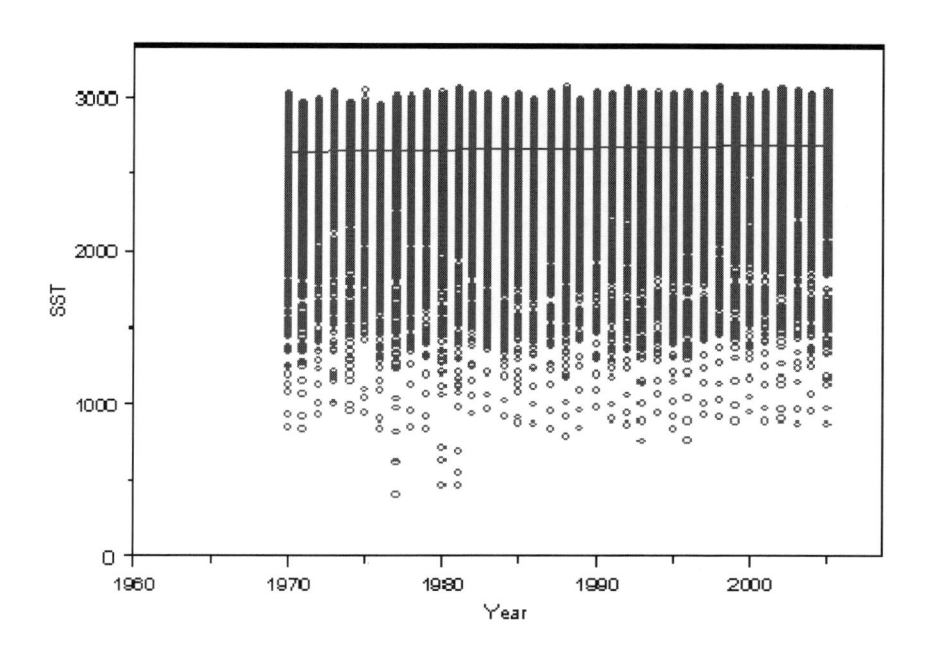

Figure 11. Box plots of SSTs over entire WNP using data for weeks 20-52.

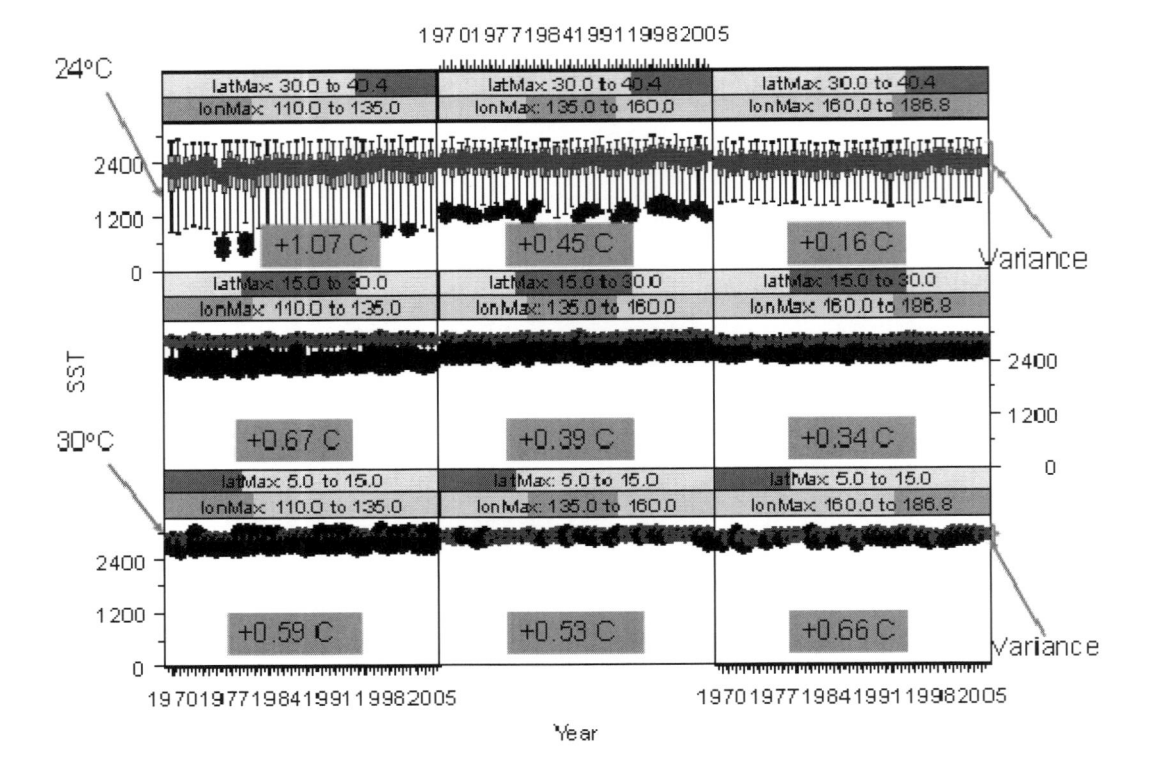

Figure 12. Conditional plot showing SSTs conditioned on latitude and longitude. Increasing trend in SST over time is evident, though different from region to region. The temperatures marked with green backgrounds are the total SST changes during 1970-2006 for the nine regions. The green background behind each regional SST change indicates that the change is considered favorable for increased TC activity.

2. Shear

Equation 4 shows a surprisingly good least squares fit for shear as a function of time and location (as evidenced by R-squared value of 0.4862).

Shear (m/s) = -16.04 + 0.0122year   (4)

where the standard errors for the intercept and the year coefficient are 9.02 and 0.0045, respectively.

This equation shows that shear, like SST, is dependent on year (our proxy for global warming), and the contribution of year is about a 0.4 m/s increase over the past 36 years. The coefficient for our year term, +0.0122 is an estimate of the relationship between year and shear, and has units of (m/s)/year. As year is our proxy for global warming, equation (4) indicates that there is a positive relationship between global warming and WNP shear. Note too, that again the coefficient scales the equation and keeps the units consistent, so the size of the coefficient is not indicative of its importance.

Figure 14 shows that the shear is near zero or negative, particularly in the tropics. Moreover, the increase in shear is most readily seen in the lowest latitudes of our region of concern. Here "increasing shear" might be misleading because the shear increase means shear is tending to be less negative, but the magnitude of shear (which is what is important to TC genesis and sustainment), is actually decreasing.

As we saw with SST, shear is changing with time, and its change is clearly regionally dependent. This supports our contention that analysis of the relationships between TC activity and the LSEFs needs to be done with data on the local level e.g., our weekly five by five blocks.

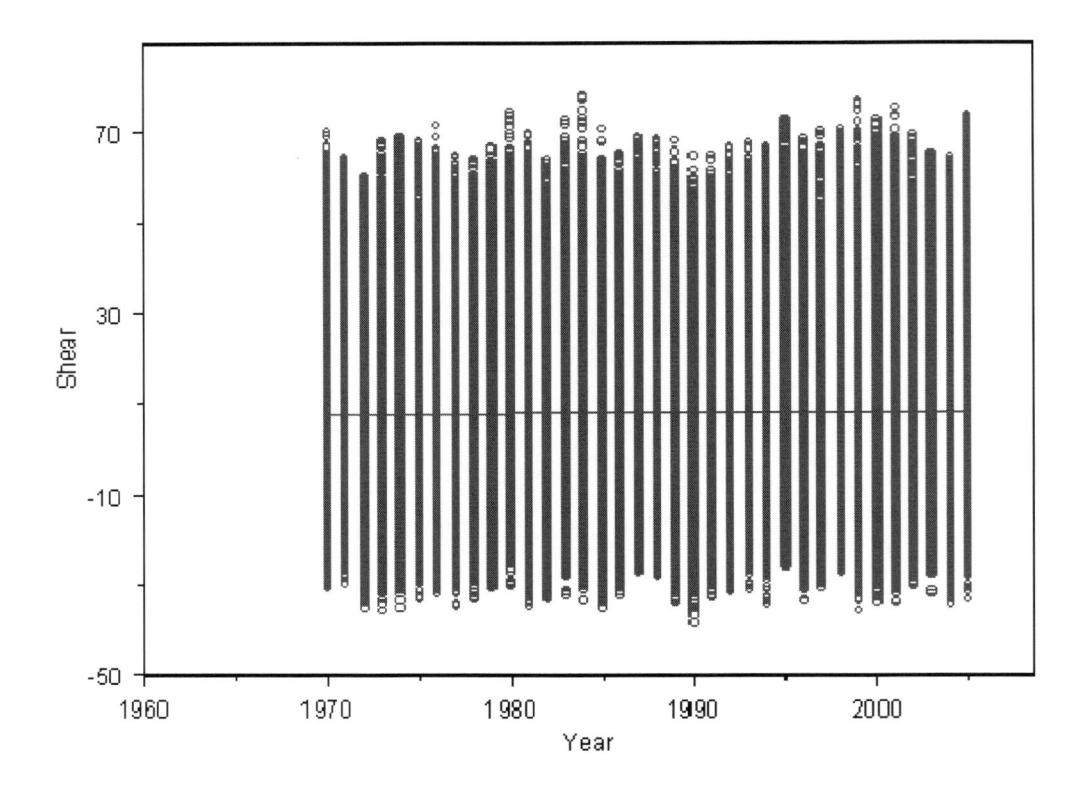

Figure 13. Least squares fit of shear data using year only as a dependant variable. Note the very slight increase in average shear with time.

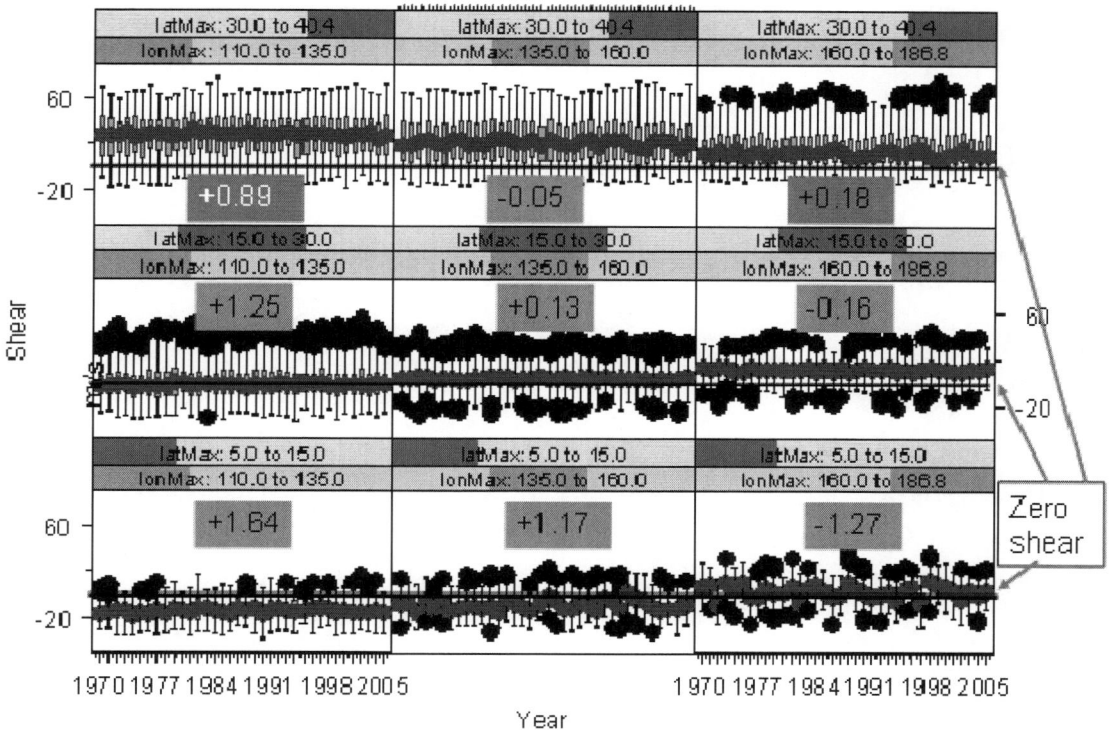

Figure 14. Shear vs. year conditioned on latitude and longitude. The change in shear (1970-2006) is shown for each region with green background indicating the change is considered favorable for increased TC activity and red considered unfavorable.

3. Vorticity

Since absolute vorticity is in part due to Coriolis effects (a function of latitude), absolute vorticity lends itself nicely to using least squares. The equation for absolute vorticity is:

$$\text{Vorticity } (s^{-1} * 10^5) = -1.100 + 0.0004 \text{year} + 0.0017 \text{week} + 0.2184 \text{maxLat} \quad (5)$$

where maxLat is the upper latitude of the 5 degree by 5 degree blocks (all data, in its geographic and weekly blocks is identified with a year, a week, a maximum latitude and a maximum longitude). The standard errors for the intercept and each of the coefficients in order are: 0.4232, 0.0002, 0.0002, and 0.0002. The value for adjusted R^2 is 0.90. Analysis of variance (ANOVA) indicates that latitude is indeed an important predictor for absolute vorticity (p-value less than 0.00001). Year, in the presence of week and latitude according to ANOVA is also a significant predictor (though barely: p-value = 0.048) whose contribution over the past 35 years has grown a negligible 0.014 s^{-1}. Figure 13 also demonstrates that absolute vorticity is predominantly a function of latitude only, and not of year, our measure of global warming.

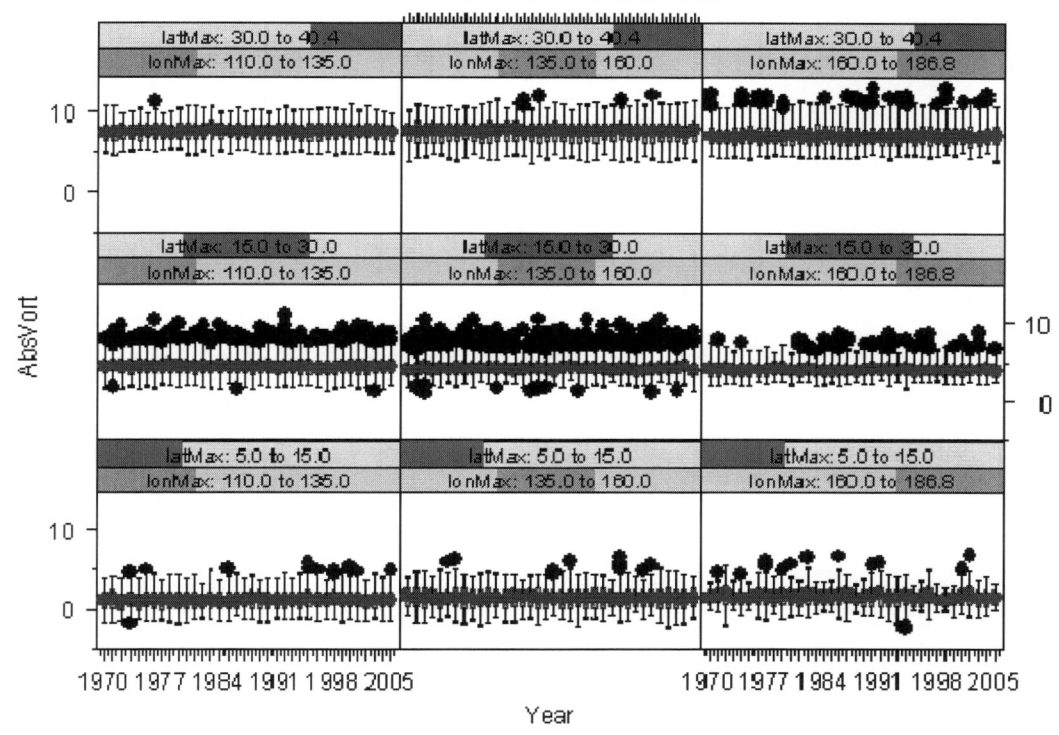

Figure 15. Absolute vorticity vs. year, conditioned on latitude and longitude. Note the predictor variable that dominates the response is latitude.

4. Mean Upward Vertical Velocity

Using a least squares fit of vertical velocity vs. year we found that there is a negligible change in vertical velocity over time, our proxy for global warming (see Figure 16).

Figure 17 also shows a lack of dependence on year in our plot conditioned on latitude and longitude. Note also, that vertical velocity changes in the transition from the Reanalysis I field (1970-1979) to Reanalysis II (1980-2006); this change is most obvious in the positive extremes.

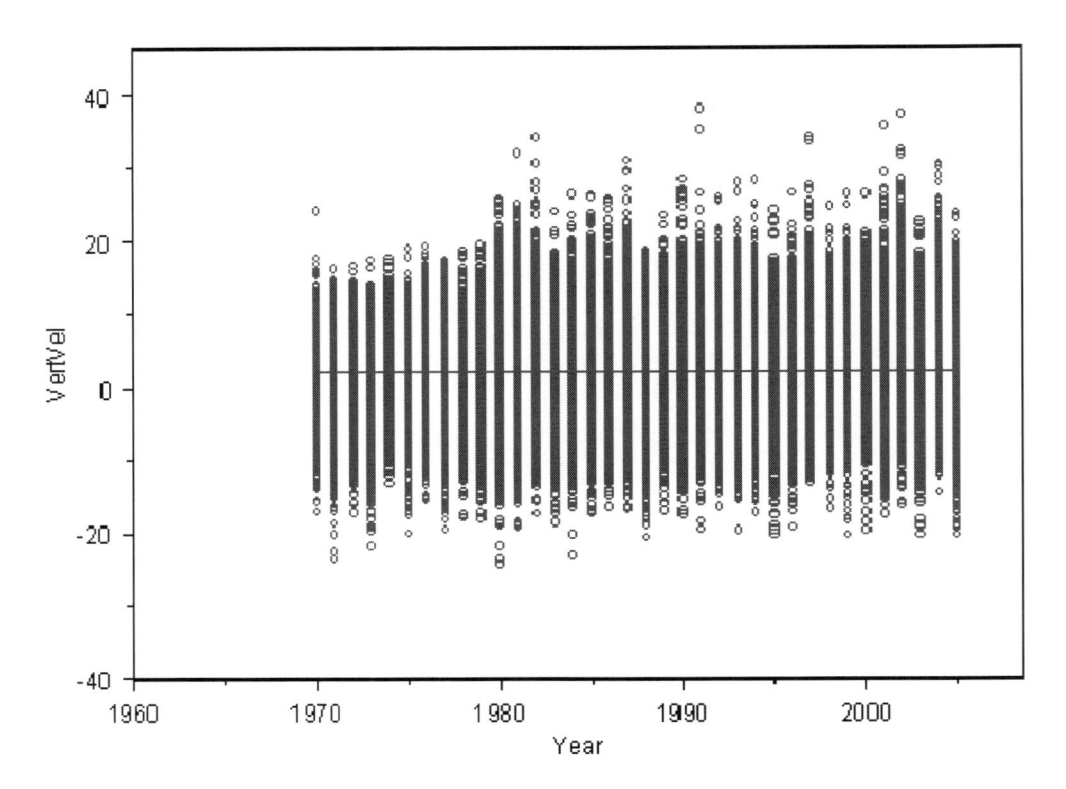

Figure 16. Least squares fit of vertical velocity vs. year.

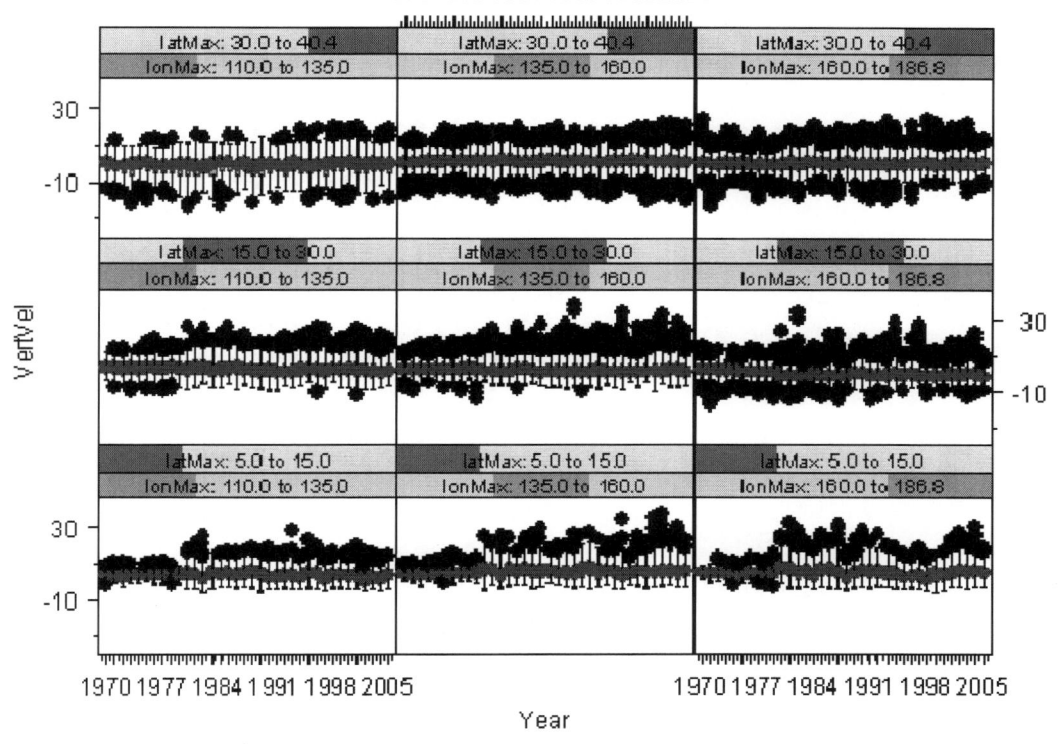

Figure 17. Box plot of vertical velocity vs. year conditioned on latitude and longitude.

5. Relative Humidity

When examining relative humidity using a simple least squares fitted line, we found that there was a significant positive trend over time (Figure 18). However, examination of Figures 19 and 20 indicates that the upward trend is artificial and due to differences between the Reanalysis I and II data sets.

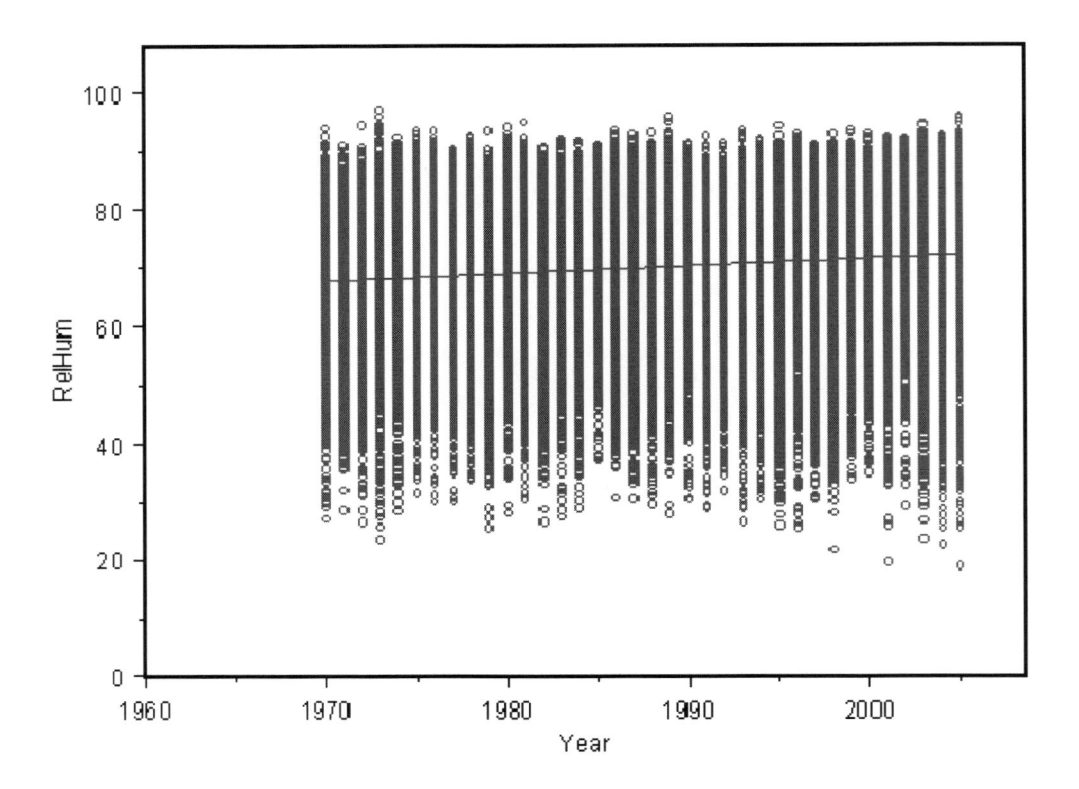

Figure 18. Least squares regression line based on both Reanalysis I and II data showing an increase in relative humidity over time.

Relative humidity, using only data from the reanalysis II data set, was analyzed with the least squares fitted line shown in Figure 21. We see that when analyzing one consistent data set, relative humidity, instead of sizably increasing over time over the WNP, decreased very slightly (a total of about 0.37 percent (compared to about 70 percent) over the past 25 years. Figure 22 indicates a seeming lack of dependence of relative humidity on latitude or longitude.

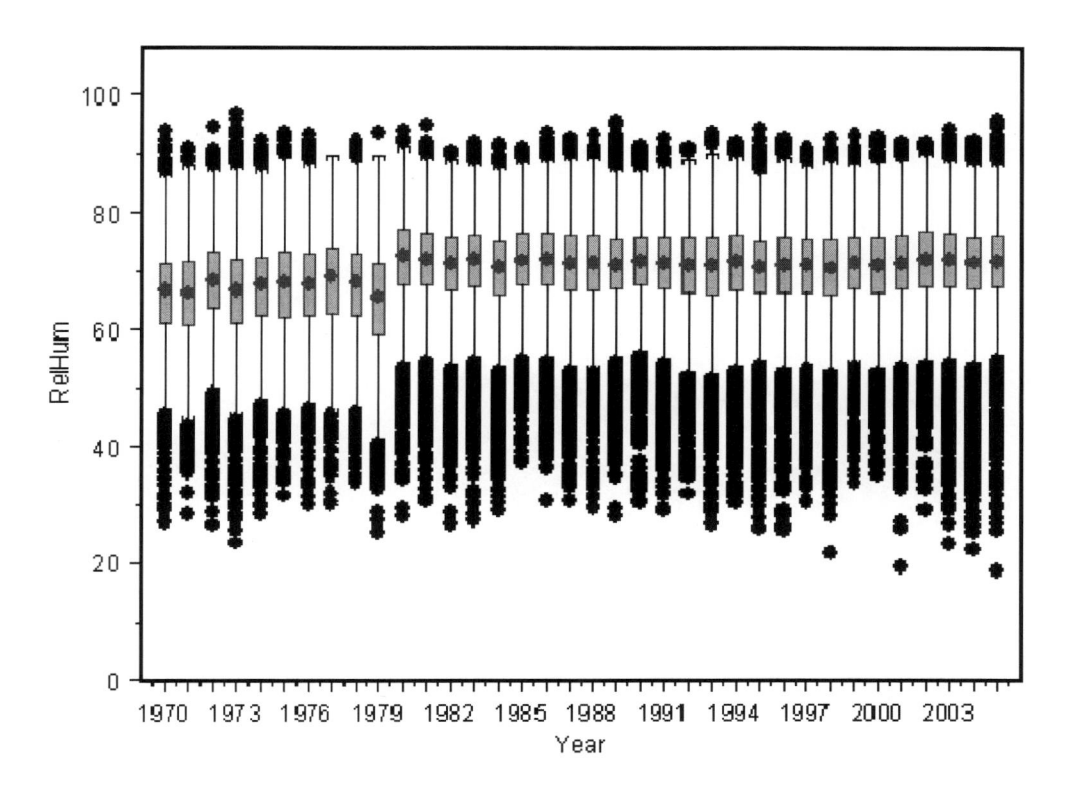

Figure 19. Box plot of relative humidity showing discontinuity at the 1979-1980 transition between the Reanalysis I and II data sets.

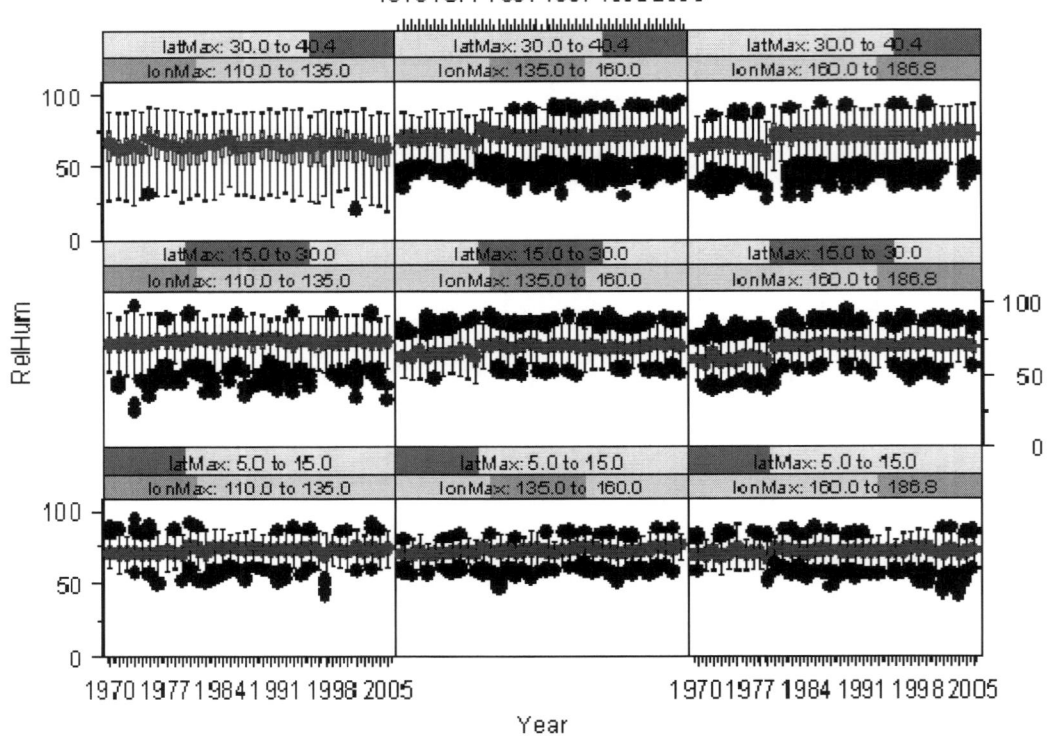

Figure 20. Box plots of relative humidity conditioned on latitude and longitude. The discontinuity between data sets at 1980 is again readily apparent.

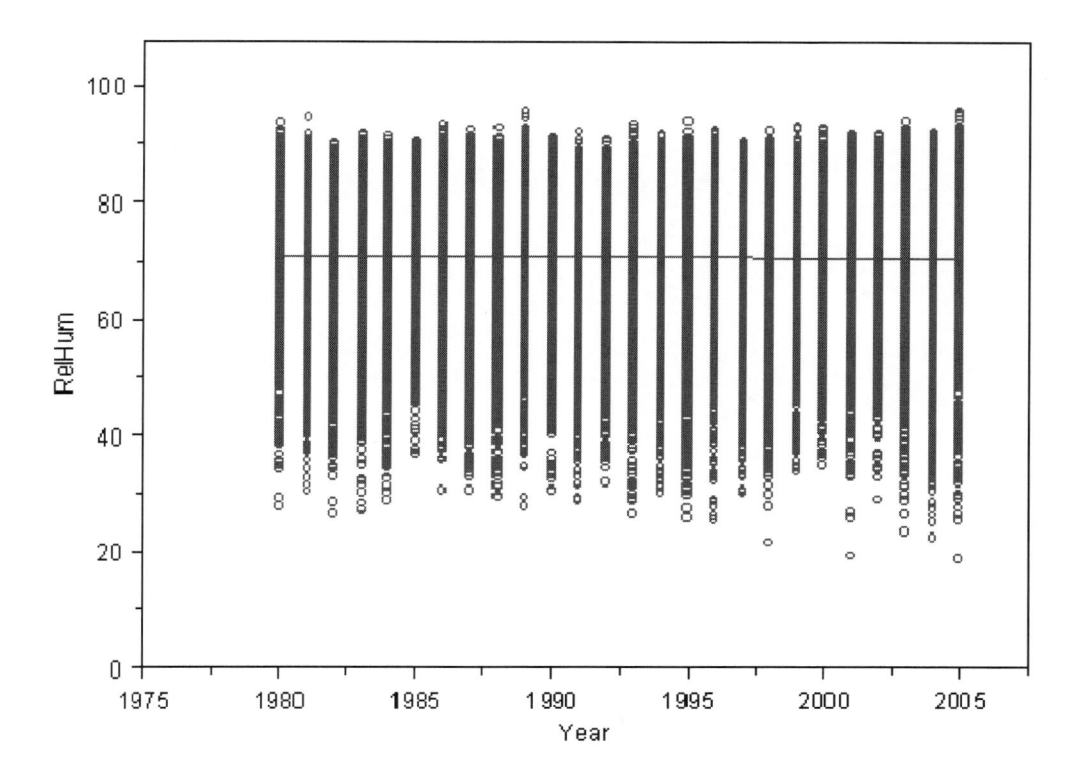

Figure 21. Plot of least squares fit using Reanalysis II relative humidity data only.

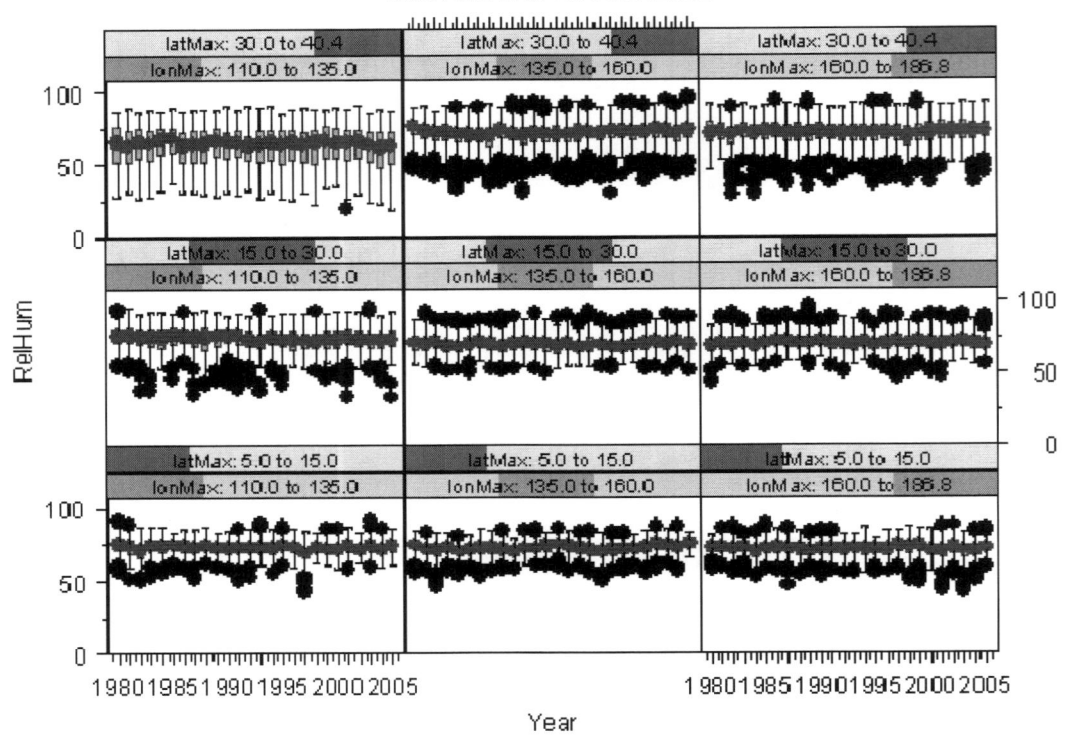

Figure 22. Relative humidity box plots conditioned over latitude and longitude.

In summary, we have found that of the LSEFs, SST and shear seem to be in part a function of year, our proxy of global warming, and that in general they have become more favorable for TC activity. Thus, SST and shear in the WNP may have been altered by global warming in a way that would favor increased TC activity. The other LSEFs do not appear to be functions of year, and thus: (a) their effects on TC activity have apparently not changed with time; and (b) they do not appear to have been altered by global warming.

B. LONG TERM TRENDS IN TROPICAL CYCLONE ACTIVITY
1. Tropical Cyclone Frequency

Figure 23 shows the least squares fitted line of TC frequency in the WNP during 1970 - 2005. Clearly there is an increasing trend with time, with the equation for this line as follows:

$$TCCount = -273.82 + 0.1529*year \qquad (6)$$

where year is considered significant as a predictor, having a p-value of 0.048. The standard errors for the intercept and the year term are 148.27 and 0.0746 respectively.

Equation 6 indicates that the average number of TCs has increased since 1970 by over five per year, again over the entire WNP. R^2 for the regression is exceptionally low (0.11). This low value is an indication that TC frequency is dependent on more than just year, but also other independent variables, such as the LSEFs.

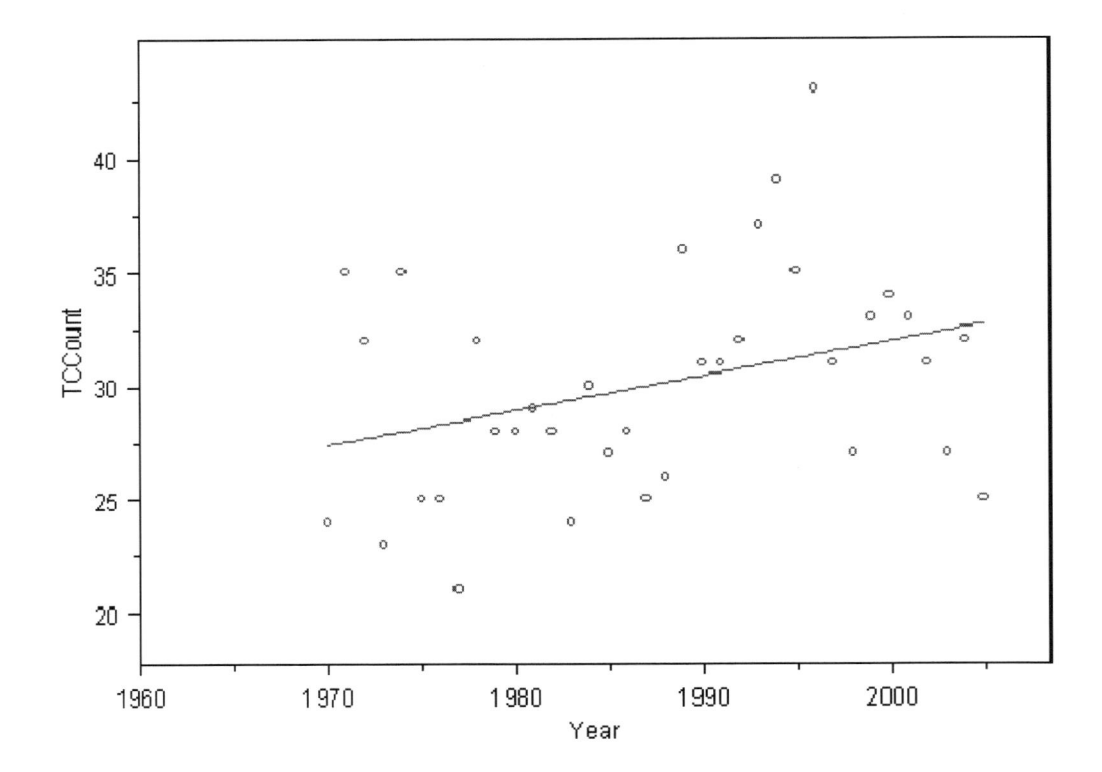

Figure 23. Least squares linear regression of TC formation for 1970-2005 in the Western North Pacific.

While there does appear to be some nonlinearity in the expected annual TC formations as a function of time, the variability of annual TC formations in Figure 23 appears relatively constant over time. The normal quantile-quantile plot of the residuals of the regression fit in Figure 24 supports that the underlying regression assumptions are reasonable for this data fit.

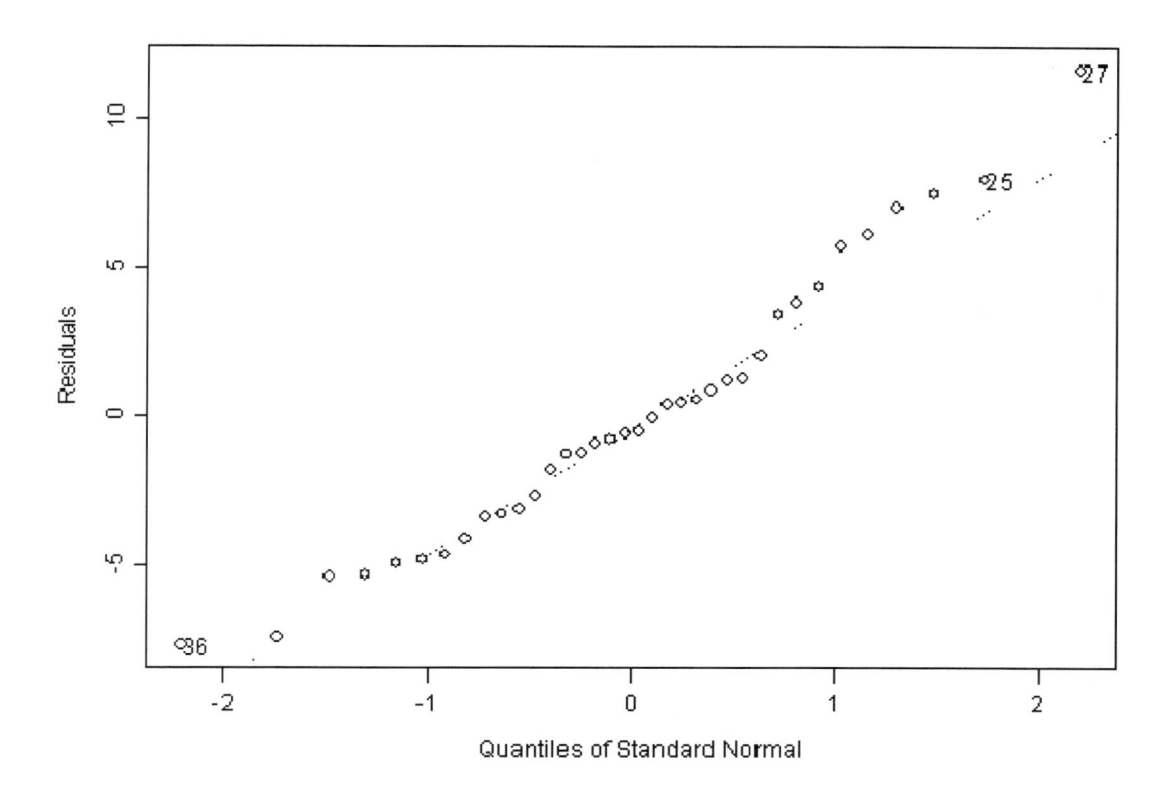

Figure 24. Normal probability plot showing a normal distribution of the residuals.

So far, we have seen that of the LSEFs, SST and shear seem to be in part a function of year, our proxy for global warming, and that for the most part they are tending in a direction which favors TC activity. We now also have Equation 6, which on a very broad analysis scale indicates that TC formations are also a positive function of our global warming proxy, year.

2. ACE

Figure 25 demonstrates a clear upward trend in ACE over time in the Western North Pacific over the 1970 through 2005 timeframe. Using linear regression we get the following equation:

$$ACE\ (kts^2) = -52027255 + 27695*year \qquad (7)$$

Equation 7 indicates that over the past 36 years, annual average ACE in the WNP has increased on average by almost one million knots^2. Figures 25 and 26 suggest that the linear regression modeling assumptions of homoschedasticity and normal errors are reasonable.

As with the results for TC frequency, the value of R-squared is low (0.083), which is again due the fact that in this model the only independent variable in Equation (7) is year, and if we wanted to predict ACE with accuracy (rather than merely demonstrate the effect of our global warming proxy on ACE) we would need to include other independent variables such as the LSEFs.

There is some slight evidence (p-value = 0.0883)) that the coefficient for year estimated in Equation (7) is not zero. Had the p-value been 0.05 or less, we would have claimed that ACE is a function of year, our proxy for global warming. Had the p-value been greater than 10% we would have stated that ACE does not appear to be a function of year. As it stands, we can only state that there may be a relationship, and try to establish what that relationship is when we examine ACE as a function of the LSEFs later in this study.

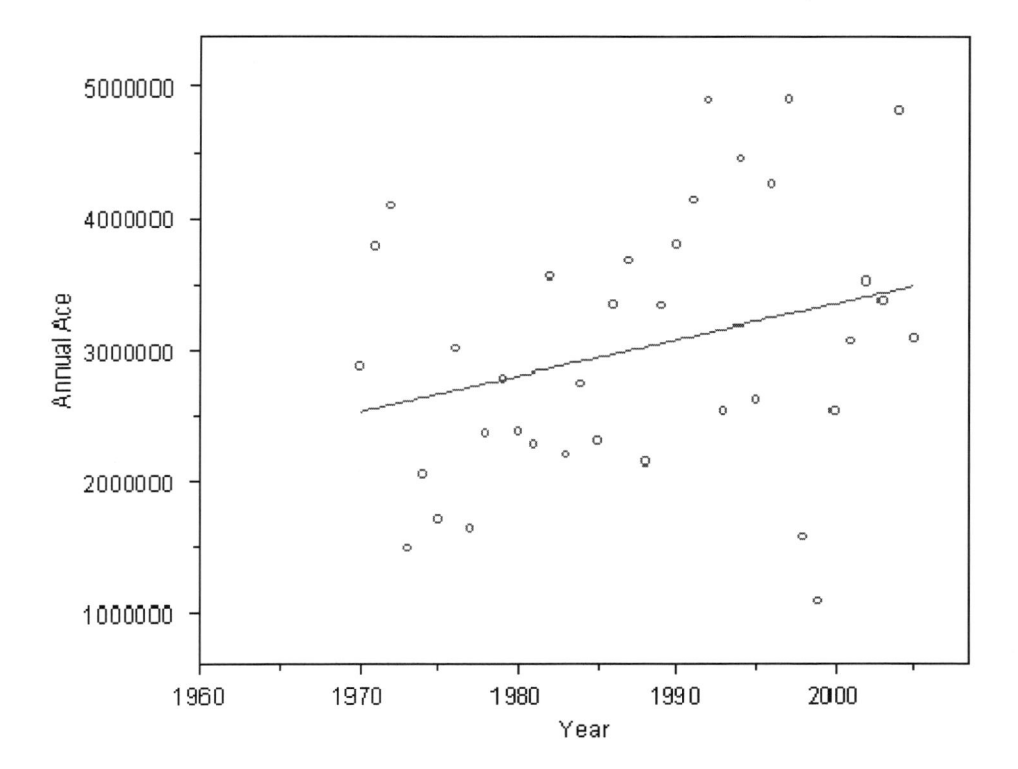

Figure 25. ACE (kt^2) vs. year for the WNP, with linear trend included.

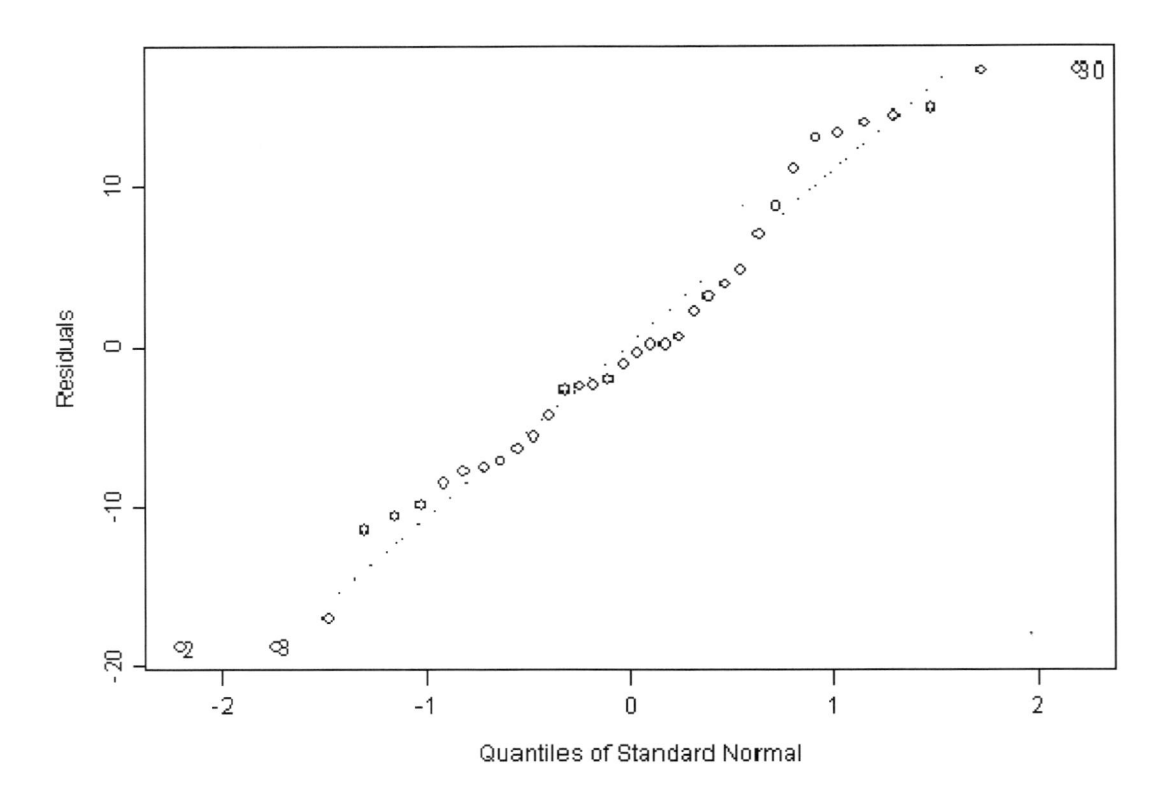

Figure 26. Normal probability plot of the ACE residuals showing the residuals are indeed normally distributed.

The ACE results are intriguing. We expect that if TC formations increase, there might be a corresponding increase in ACE. However, Figure 27, which is a plot of ACE per storm for each year, shows an increase in ACE over time even though this average should show a trend line that hovers around zero if the increase in ACE was due solely to an increased number of storms. This is consistent with other studies[39] which suggest that the intensity of individual TCs has been increasing over time.

We have shown that of the LSEFs, SST and shear seem to be in part a function of year, our proxy of global warming,

---

[39] P. J. Webster, et al., *Science*, **309**, 1844 (2005).

and that for the most part they are tending in a direction which favors TC activity. We also have Equation 4, which on a broad scale indicates that TC formations are also a positive function of our global warming proxy, year, and Equation 7, which on a broad scale indicates that ACE is increasing. Furthermore, it appears that ACE is increasing more than can be accounted for by the increase in the number of TCs.

These results suggest (but do not prove) that: (1) global warming during 1970-2005 may have affected two of the LSEFs in the WNP, SST and shear, and (2) global warming may have contributed to an increase in TC frequency and ACE during 1970-2005. In addition, the changes in SST and shear are consistent with the TC frequency and ACE changes, indicating (but not proving) that global warming induced changes in SST and shear may have contributed to the increase in TC frequency and ACE during this period.

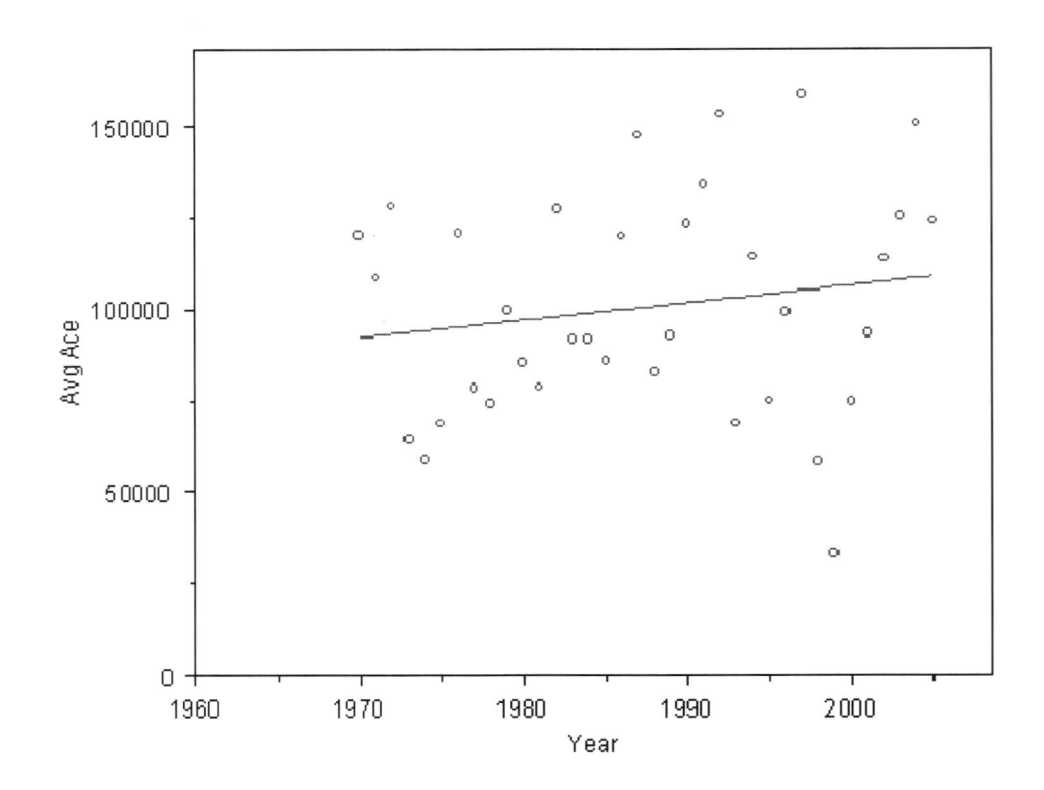

Figure 27. Average ACE per storm vs. year showing a small but pronounced upward trend.

C. MODELLING OF TROPICAL CYCLONE ACTIVITY IN THE WESTERN NORTH PACIFIC USING LARGE SCALE ENVIRONMENTAL FACTORS

1. Tropical Cyclone Frequency

Using Equation 7, we have shown a relationship between year and the number of TCs formed. Based on using the total number of TCs per year in the WNP, 1970-2005, and ignoring the LSEFs. We have also shown that there is a relationship between global warming and two of the LSEFs. We now set about modeling TC formation where the independent variables are the LSEFs, the factors known to influence TC formations.

The variable used for TC formation is binary, that is either a TC formed in our weekly five by five degree block, or a TC did not form. Logistic regression was used as our modeling tool to estimate the probability of TC formation p as a function of the independent variables. We started simply, using SST as the only independent variable.

One problem that immediately became apparent was that the model is significantly underdispersed. Underdispersion is indicated by the fact that the residual deviance should be approximately equal to its number of degrees of freedom[40]; our results show a residual deviance of 10652.02 on 144934 degrees of freedom. This underdispersion indicates that the variability of the response variable is much less than $p(1-p)$, the variance of the Bernoulli distribution assumed in logistic regression.

The cause of underdispersion might be explained because in logistic regression observations are assumed to be independent, whereas these may not always be. For example TC formation in a region means that some relatively slowly changing and wide spread influences on TC formation (such as SST) may favor the formation of other TCs. Conversely the formation and growth of a TC may alter conditions (such as shear or SST) so as to decrease the likelihood of TC formation in the immediate future and in the immediate proximity. Our conjecture is that this altering of conditions preventing other TCs from forming would act only through the LSEFs, and minimize this dependency problem.

---

[40] D. Collett, 1991, Modelling Binary Data. Chap. 5, *Overdispersion*. Chapman and Hall, 2-6 Boundary Row, London SE1 8HN, UK, 188.

With this underdispersion, caution must be taken when interpreting inference results based on the logistic regression fit to this data. However, the logistic regression fit to the data can still be used to estimate the probability of TC formation. Whether the model fits the data or not will be assessed by comparing how well estimated probabilities of TC formation compare to actual formations of an independent set of data.

In a logistic regression model, the probability of success, p (in this case "success" is the formation of a TC) is linked to k independent variables $x_1, x_2,...x_k$ as follows:

$$\text{logit}(p) = \beta_0 + \beta_1 x_1 + \beta_2 x_2 + ... + \beta_k x_k \tag{8}$$

where $\text{logit}(p)$, which is $\log(p/(1-p))$, is the log-odds of the success probability[41], and the $\beta_0,...\beta_k$ are coefficients of the linear combination of the k different independent variables.

In its simplest version (the null model), the success probability is a function only of a constant term, β_0. For progressively more complex models, the success probability is a function not only of the β_0 term, but of one or more independent variables and their corresponding coefficients.

Using a combination of variable selection techniques and diagnostic plots gives the model fit summarized in Table 3.

---

[41] D. Collett, 1991, Modelling Binary Data. Chap. 5, *Overdispersion*. Chapman and Hall, 2-6 Boundary Row, London SE1 8HN, UK, 188.

Term	Coefficient	Standard Error
Intercept	-22.3677	2.3205
SST	0.0066	0.0007
Coriolis	1.3588	0.1954
AbsVort	0.0407	0.0254
VertVel	0.1436	0.0095
RelHum	-0.0276	0.0097
Shear	-0.0264	0.0060
Shear^2	-0.0031	0.0004

Table 3. Estimated coefficients and standard errors for model to predict the formation of a TC in a weekly five by five degree block.

In the course of model development and variable selection we learned that our model was very sensitive to the changes between the two NCEP reanalysis data sets. As the majority of our data was from the better Reanalysis II data set, we used only the Reanalysis II data set (1980-2006) in the development of the model. At the same time, we reduced the data set, retaining only the weekly five by five degree blocks with TC formations or ACE, as well as a randomly selected twenty percent of the remaining TC activity free blocks. This was done to reduce computer runtime and memory issues, and it also reduced, but did not eliminate our underdispersion problem.

Table 3 shows that we have included a Coriolis term in our model. Coriolis effects are an implicit large scale factor that is necessary for TC formation[42], with values near zero (e.g., within a few degrees of the equator) being unfavorable for TC activity. The Coriolis effect is a strong function of latitude and is unlikely to change significantly with global warming. However, when running the above model without a Coriolis term, the result was significant overprediction at and close to the equator. Thus, we included a Coriolis term in the model.

Care must be taken when interpreting the magnitude of the estimated coefficients given in Table 3. These coefficients represent the partial effect of each predictor in the presence of the rest. This is more understandable when you examine Figure 28. There are noticeable, though not terribly strong relationships between mid-level relative humidity and in particular, SST and vertical velocity. These relationships obscure the contribution to TC formation by relative humidity, because, if you have information on SST for example, you also have some information on relative humidity.

---

[42] W. M. Gray, 1975: *Tropical Cyclone Genesis in the Western North Pacific*, Naval Air Systems Command, Washington DC, 20361.

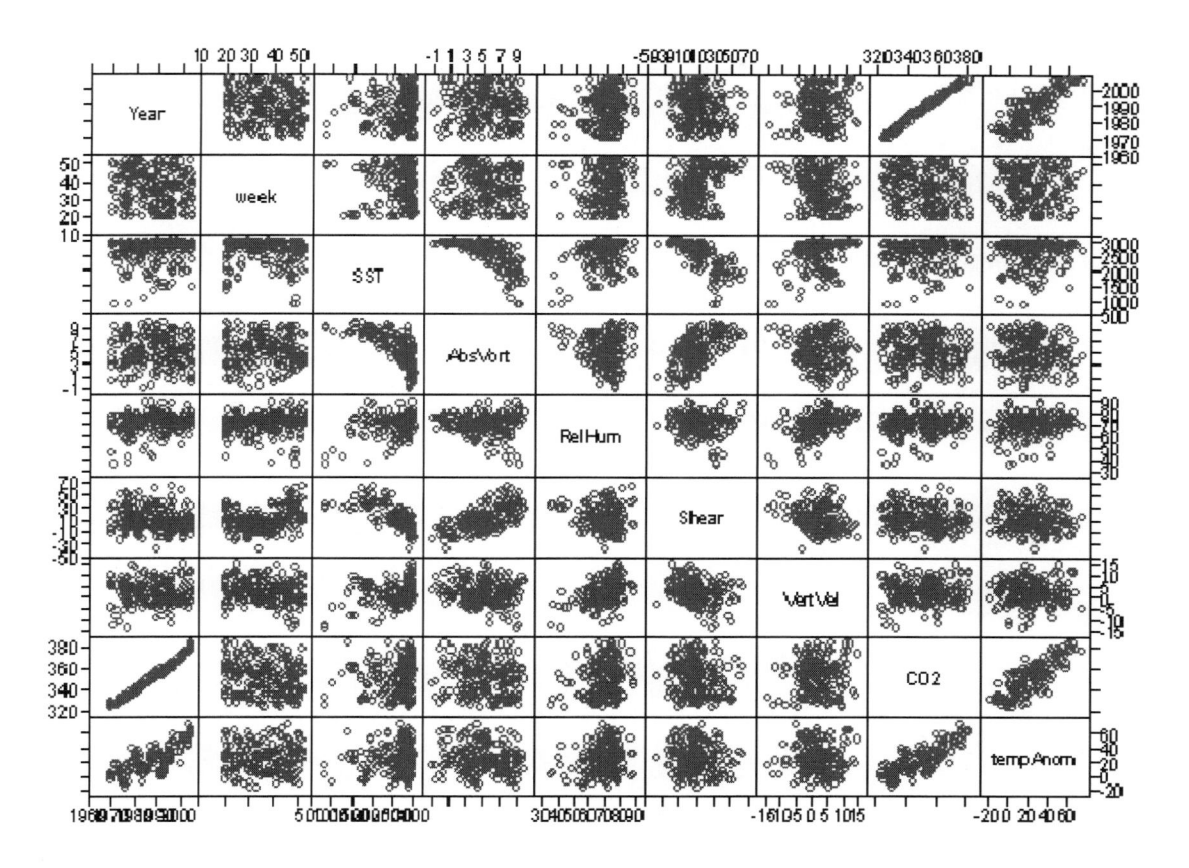

Figure 28. Scatter matrix of time, main factors, CO_2 concentration, and temperature anomaly.

Table 3 also indicates that among the independent variables, a quadratic term of shear is also included, and is significant. Polynomial terms are necessary to model the log-odds as a non-linear function of shear. While shear ranges from approximately -40 m/s to 70 m/s, it is most conducive to TC activity at near zero values. A graph showing the contribution of the shear terms to the estimated log-odds is seen in Figure 29. The slight left skew of the shear contribution may be due to some physical phenomenon, but most likely is due to the fact that most TCs form in or near the tropics where the average shear is negative. It is no surprise that the model with the best fit demonstrates that the contribution of shear to the probability of TC formation is zero when shear is near zero, and becomes sharply negative as the magnitude of shear increases.

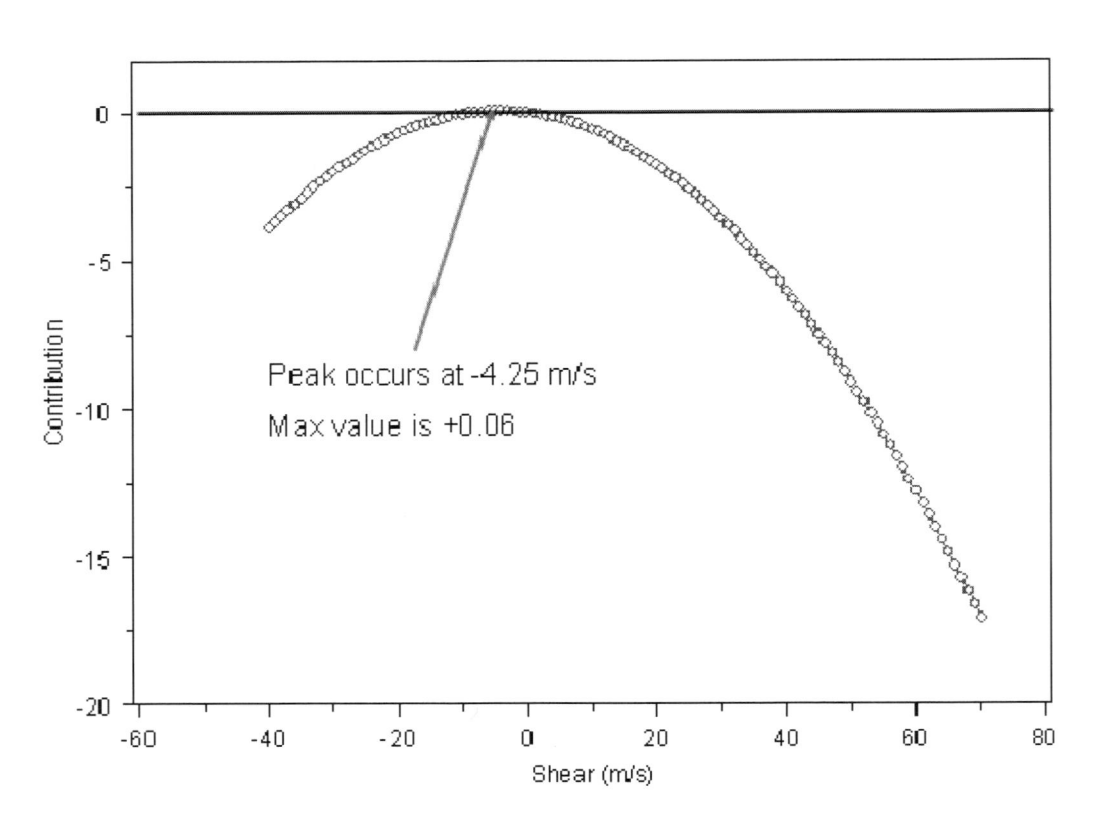

Peak occurs at -4.25 m/s
Max value is +0.06

Figure 29. Contribution to TC formation probability model by the main effect shear.

The one obvious remaining issue is that the model of Table 3 indicates a negative relationship between relative humidity and TC formation, the opposite of what was expected. However, Figure 30 helps explains this issue.

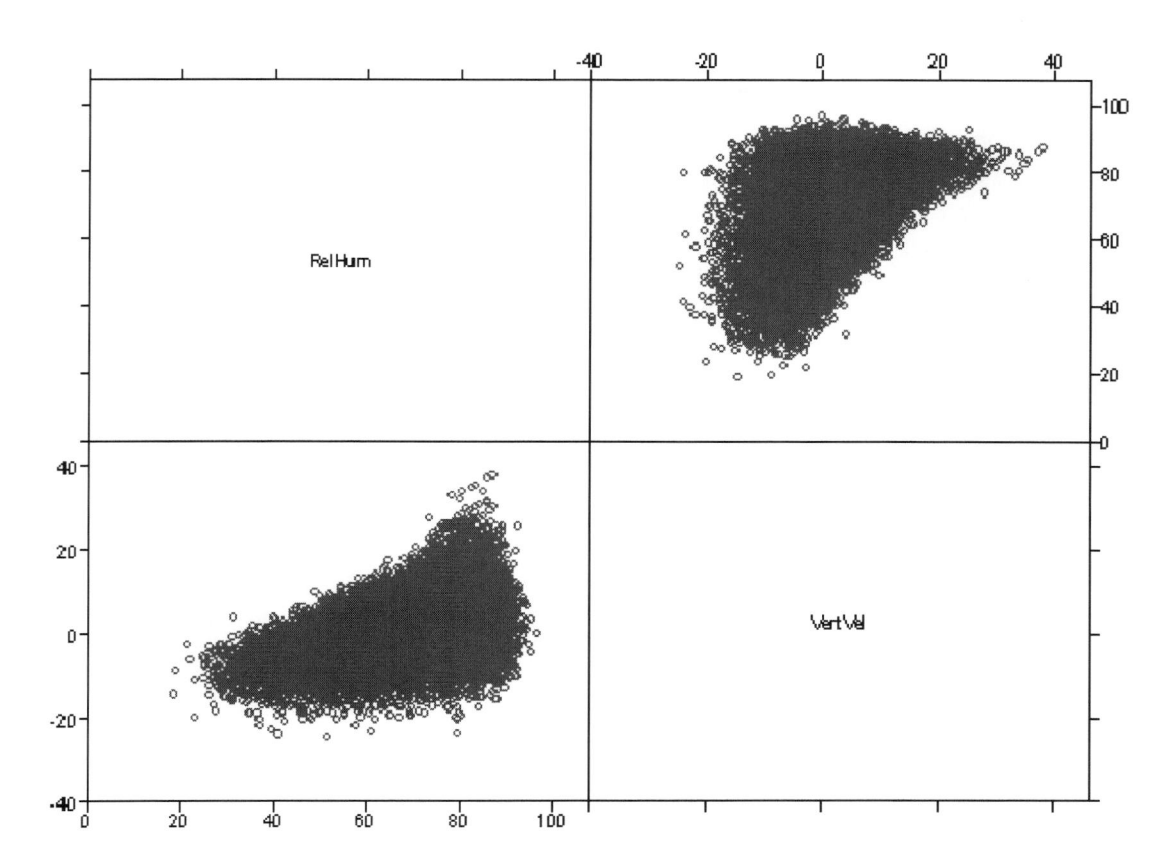

Figure 30. Scatter matrix of relative humidity and vertical velocity.

Figure 30 shows there is a strong positive relationship or multi-colinearity between relative humidity and vertical velocity. This multi-colinearity, combined with the knowledge that high relative humidities are required for TC formation and that vertical velocity has an upper bound on its contribution to TC formation, causes the coefficient for our relative humidity term to be weakly negative, even though the marginal contribution of relative humidity to TC formation is most definitely positive. The residual plots in the next few sections help shed light on this behavior.

The end result of Table 3 is that if we want to characterize the contribution of each of the LSEFs to TC formation, then an increase in SST, absolute vorticity, and/or vertical velocity --- or a reduction in the magnitude of shear --- will result in an increase in TC formation probability in the presence of the other factors.

A more intuitive feel for the relationship between TC formation and the LSEFs can be obtained by examining the partial residual plots for each of the factors. Partial residual plots are used as (among other things) diagnostics to check the form of the independent variables. In this case each of the relationships between the response and an independent variable appears linear (except for having to account for crossing through zero with shear as explained earlier), and so no further transformations of the data were required.

a. The Effect of SST on TC Formation

Figure 31 shows the partial residual plot for SST against TC formation. To aid the reader who is unfamiliar with partial residual plots, in Figure 31, the lower "line" of data represents data points for which a TC did not form for each weekly five by five degree block (at the specified SST ($^{\circ}C*100$)). The upper line is comprised of all the weekly blocks at the specified temperature where a TC did form. The vertical axis is a measure of the partial effects of SST on the log-odds that a TC will form.

There are approximately 1000 TC formations represented in Figure 31. Notice that perhaps one TC out of 1000 formed in water for which the temperature was less than 26 $^{\circ}$C. Clearly, as you examine the figure from left to right, examining the density of the data points, it is obvious that the warmer the water is, the more likely it is that a TC will form. The residual plot of Figure 31 allows us to draw some important conclusions:

1. The LSEF identified in Chapter I, sea surface temperatures (SSTs) exceeding 26 °C[43], seems to be confirmed by our residual plot.
2. Increasing SST increases the partial effect of SST on the log-odds that a TC will form.
3. We have seen earlier that global warming appears to have a positive relationship with SST. Now we observe that increasing SSTs appear to have a positive relationship with TC formation.

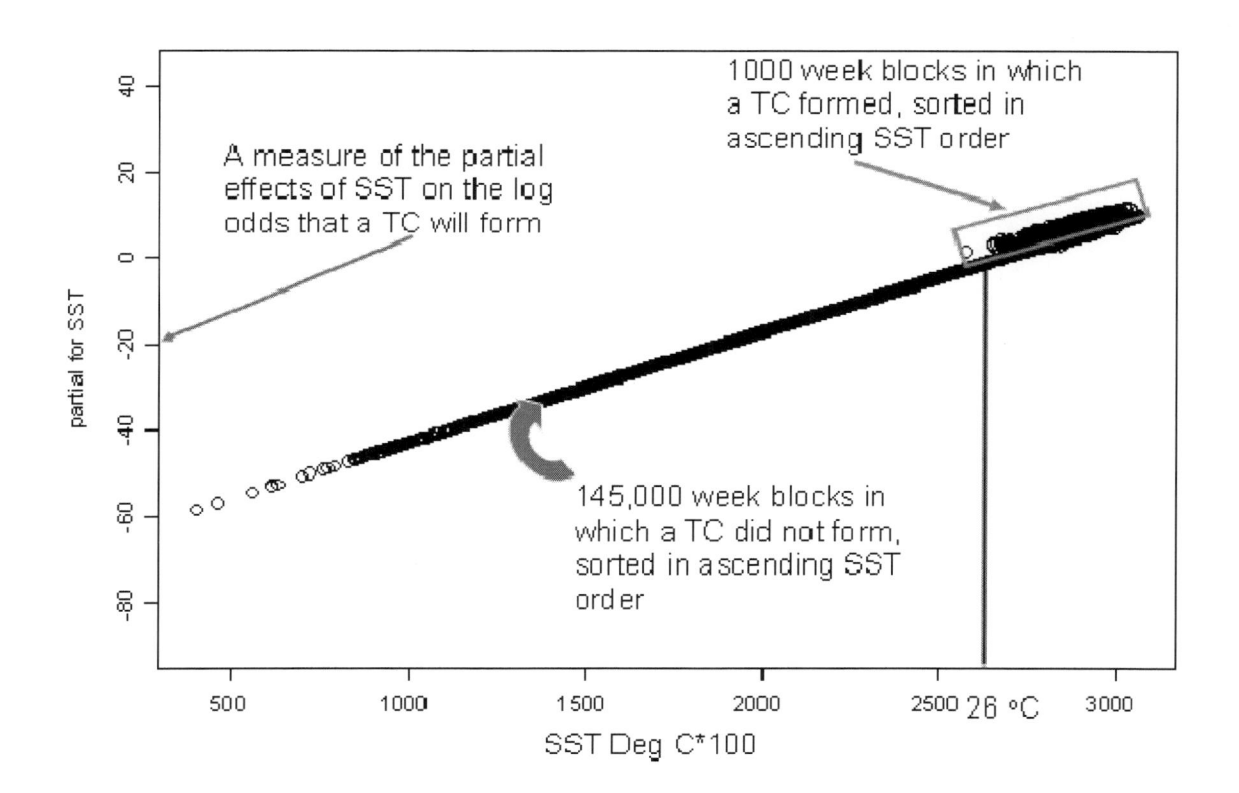

Figure 31. Partial residual plot for SST as used in our logistic regression model.

---

[43] W. M. Gray, 1975: *Tropical Cyclone Genesis in the Western North Pacific*, Naval Air Systems Command, Washington DC, 20361.

b. The Effect of Shear on TC Formation

Figure 32 is the partial residual plot for our shear term. Again, a clear pattern can be observed. Out of over one thousand tropical cyclones examined, fewer than ten formed in conditions where the magnitude of the shear exceeded 20 m/s. The vast majority of TCs which form do so where the magnitude of the shear is less than 20 m/s, with the greatest concentration occurring exactly where one would expect, where shear is near zero.

The residual plot of Figure 32 allows us to draw some important conclusions:

1. The LSEF identified in Chapter I, weak vertical shear of the horizontal winds[44], seems to be confirmed by our residual plot.

2. The occurrence of vertical shear with a magnitude greater than 20 m/s means it is very unlikely a TC will form.

3. We have shown in prior sections that for the WNP, and based on 1970-2005 data: (a) global warming appears to have a positive relationship with TC frequency; (b) global warming appears to have a positive relationship with shear; (c) global warming appears to have a positive relationship with shear magnitude. These results, combined with the positive relationship between low shear magnitude and TC frequency (Figure 32) indicates that the apparent impacts of global warming on TC frequency in the WNP have occurred via the effects of global warming on shear.

---

[44] W. M. Gray, 1975: *Tropical Cyclone Genesis in the Western North Pacific*, Naval Air Systems Command, Washington DC, 20361.

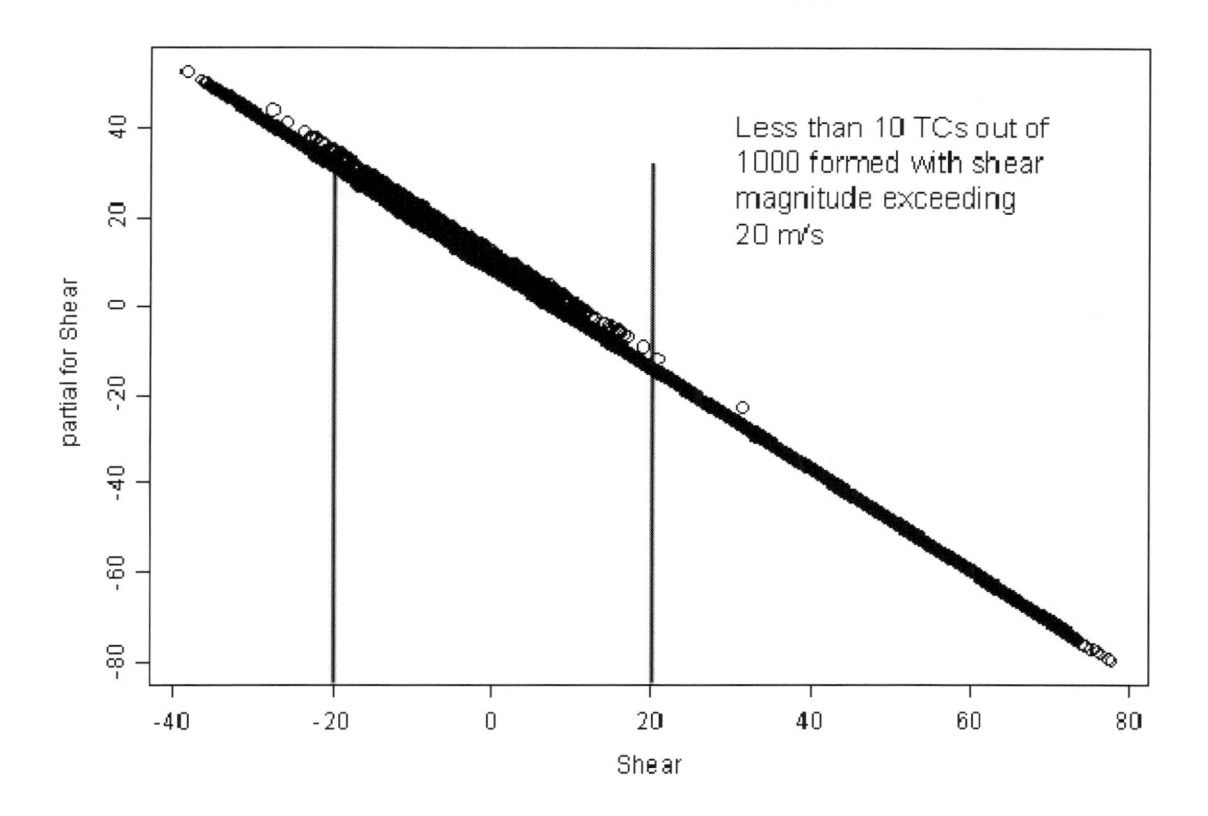

Figure 32. Partial residual plot for shear as used in our logistic regression model.

c. *The Effect of Relative Humidity on TC Formation*

We have hypothesized that relative humidity plays a role in TC formation. The partial residual plot of Figure 33 bears this out, as almost all TCs that have formed in the Western North Pacific since 1970 did so when the relative humidity was 55% or greater. Certainly Figure 33 demonstrates that high relative humidity is positively associated with TC formation. The greatest concentration of TC formations occurs at what appears to also be the greatest concentration of relative humidities (approximately 75 to 80%). This suggests that relative humidity plays a threshold role; that is, once relative humidity exceeds a certain limit, perhaps 55%, the relative humidity is favorable to TC formation, but does not further increase the formation probability as humidity continues to climb. Clearly Figure 33 confirms the LSEF: high mid-level humidity[45].

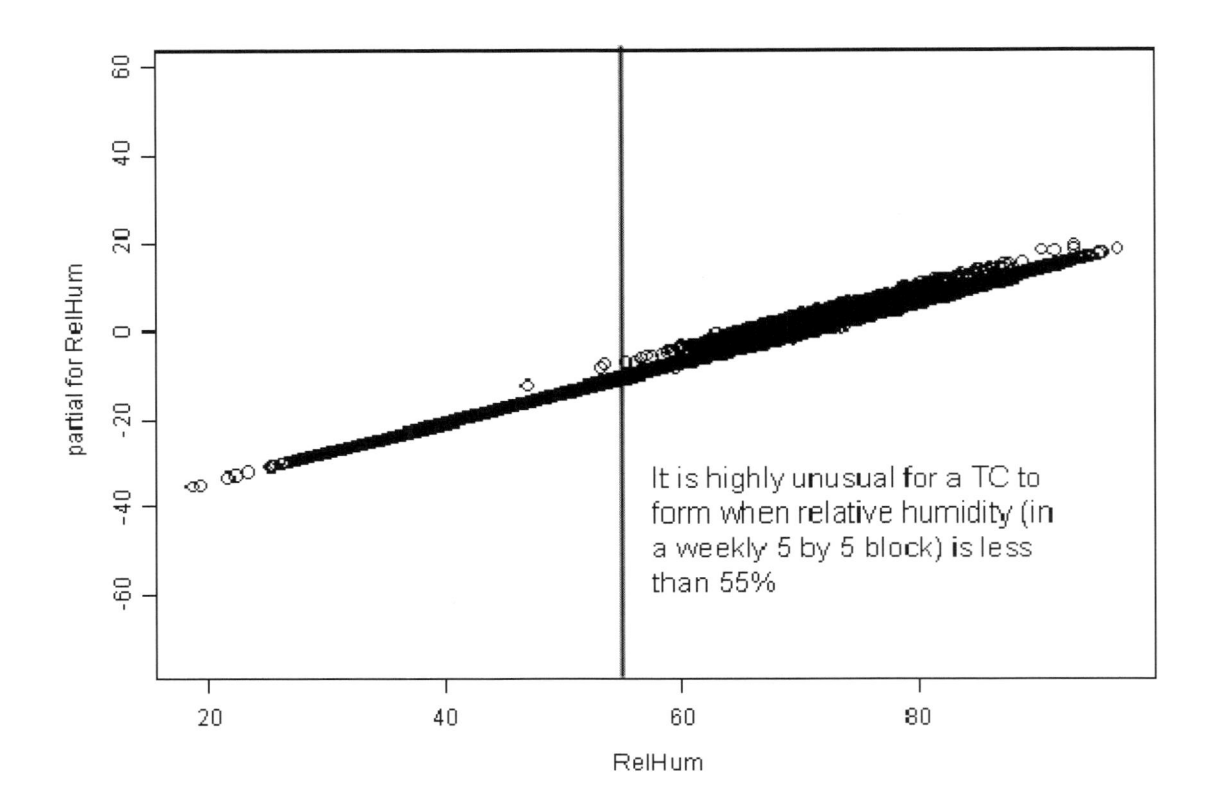

Figure 33. Partial residual plot for relative humidity as used in our logistic regression model.

---

[45] W. M. Gray, 1975: *Tropical Cyclone Genesis in the Western North Pacific*, Naval Air Systems Command, Washington DC, 20361.

d. The Effect of Vertical Velocity on TC Formation

For this discussion of vertical velocities, recall from Chapter II that we multiplied the NCEP reanalysis vertical velocities (in (Pa/s)) by -1 prior to analyzing them. Thus, in our results, positive vertical velocities correspond to upward motion of the atmosphere.

Figure 34 shows a clear relationship between vertical velocity and TC formation. Not one TC formed in conditions where the vertical velocity was below -5 Pa/s and few formed above approximately 25 Pa/s. Thus, the LSEF mean upward velocity[46] is borne out by Figure 34.

---

[46] W. M. Gray, 1975: *Tropical Cyclone Genesis in the Western North Pacific*, Naval Air Systems Command, Washington DC, 20361.

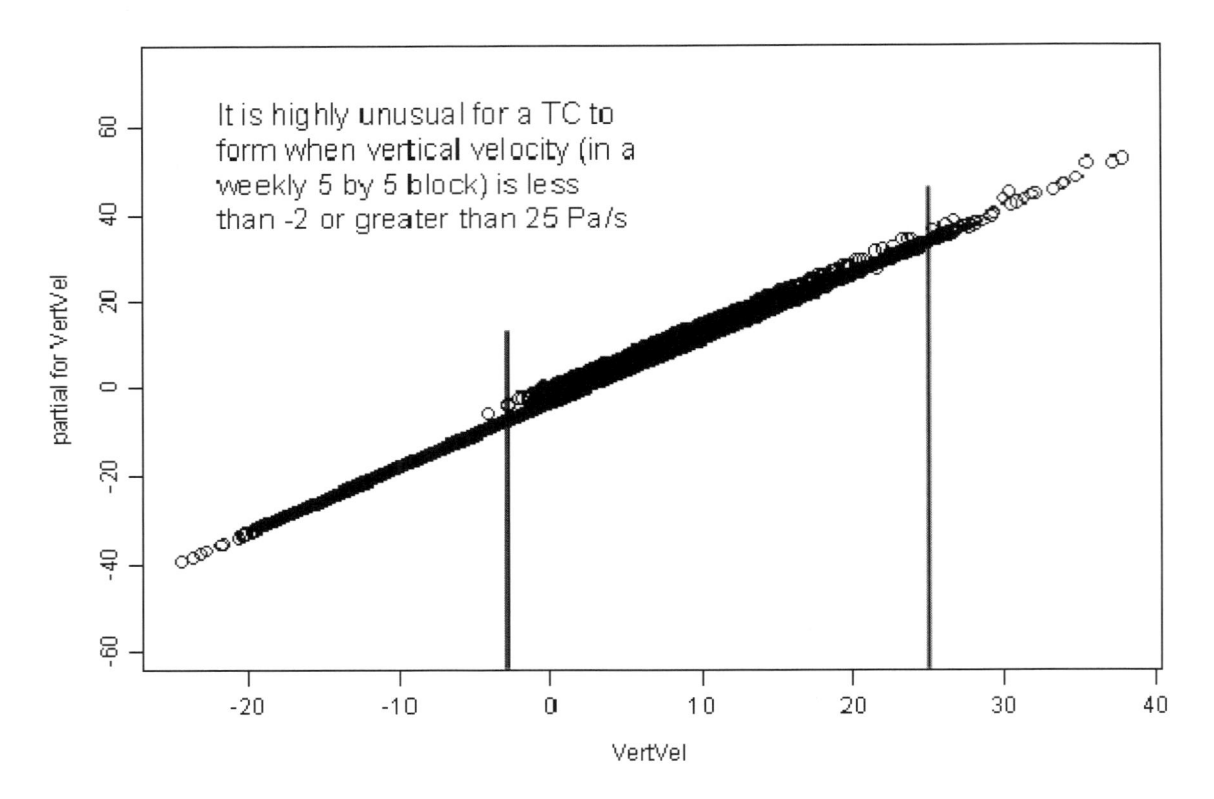

Figure 34. Partial residual plot for vertical velocity as used in our logistic regression model.

e. The Effect of Absolute Vorticity on TC Formation

We hypothesized that absolute vorticity is significant as a factor which influences TC activity. Figure 35 helps confirm the LSEF, large positive absolute vorticity at low levels[47], as it shows that virtually all TCs have formed when absolute vorticity was greater than zero. Of note, it appears that the concentration of TC formations decreases sharply at approximately 9 $s^{-1}10^5$. Absolute vorticity has a very strong relationship to latitude, and most likely the decline in the concentration

---

[47] W. M. Gray, 1975: *Tropical Cyclone Genesis in the Western North Pacific*, Naval Air Systems Command, Washington DC, 20361.

of TC formations with higher absolute vorticity is actually due to the fact that higher vorticity is associated with higher latitude and thus, lower SST and higher shear.

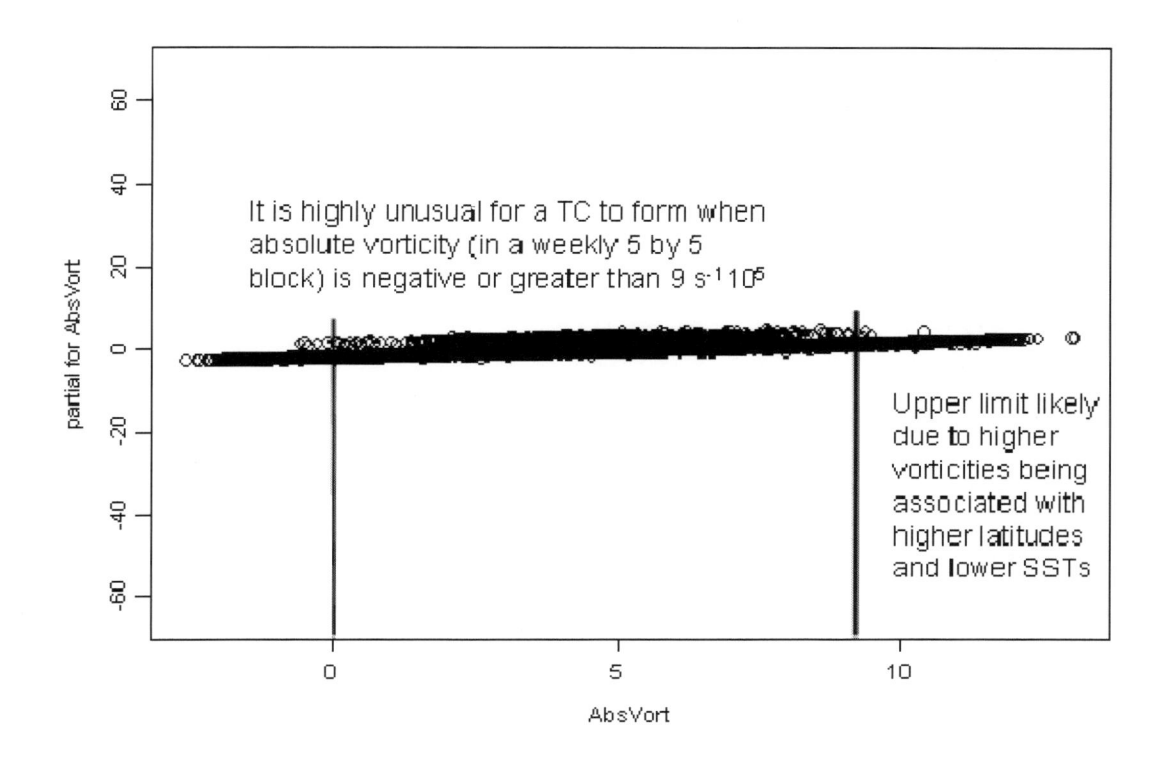

Figure 35. Partial residual plot for absolute vorticity as used in our logistic regression model.

f. Model Validation

Thus, the model of Table 3 leads to the equation below which is used to estimate the probability of TC formation:

$$\text{logit}(p) = -22.367 + 0.0066\text{SST} + 1.359\text{Coriolis}$$
$$+ 0.0407\text{AbsVort} + 0.1436\text{VertVel}$$
$$- 0.0276\text{RelHum} - 0.0264\text{Shear}$$
$$- 0.0031\text{Shear}^2 \tag{9}$$

All the terms in Equation 9 are physically reasonable, given the explanation of the negative sign for the relative humidity term presented previously. However, as stated earlier when we first discussed the underdispersion of our model, we must demonstrate model validity by how well it estimates the probability of TC formation. We constructed contour plots of estimated probability for each of our weeks included in the data base for the years 1980 through 2005. If the model is valid, it ought to be able to accurately map out regions of the WNP with relatively high TC formation probabilities. By plotting actual TC formation locations for each week on the probability contour plot for that week, we can assess the accuracy of the model probability maps. Figures 36-38 show examples of model weekly TC formation probability maps along with actual TC formation locations. Note the resemblance of the higher probability regions to regions known to be generally favorable for TC formation (e.g., the monsoon trough at about 5-15N).

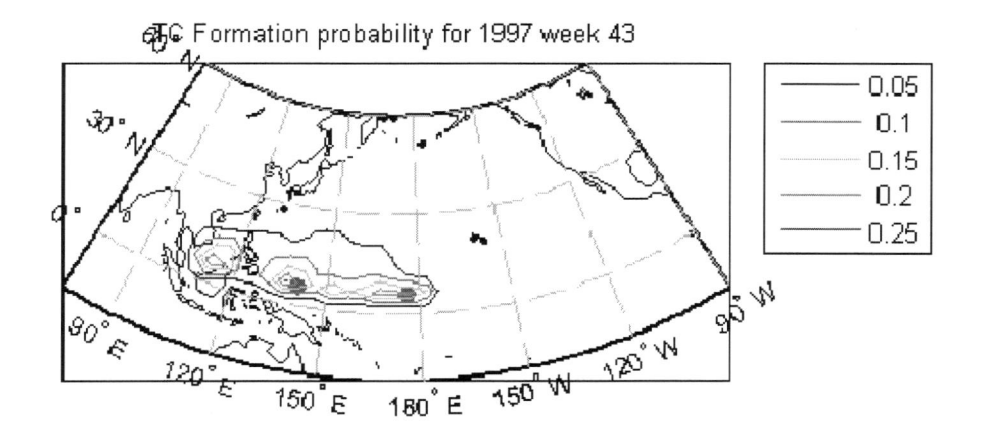

Figure 36. Model weekly probability contours for TC formation for week 43 of 1997. Red dots mark formation locations for actual TCs.

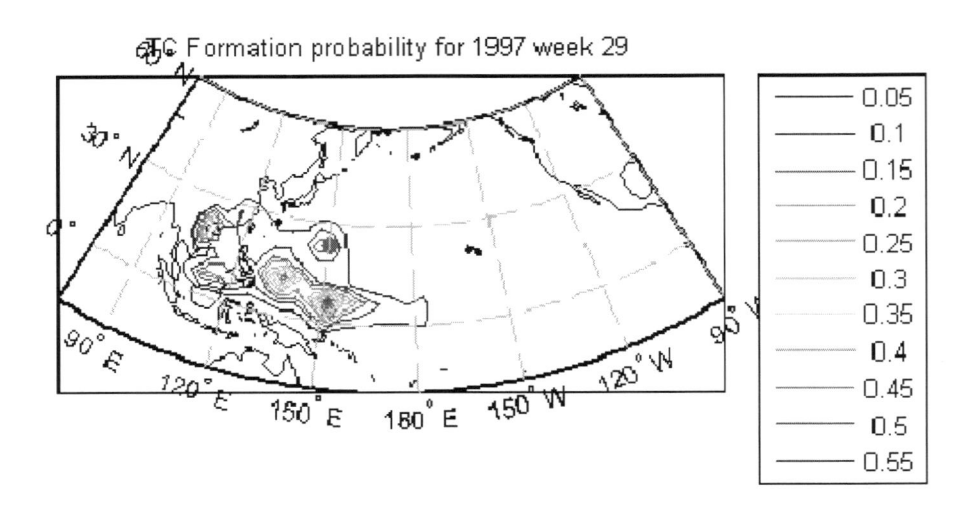

Figure 37. Model weekly probability contours for TC formation for week 29 of 1997. Red dots mark formation locations for actual TCs.

TC Formation probability for 1997 week 50

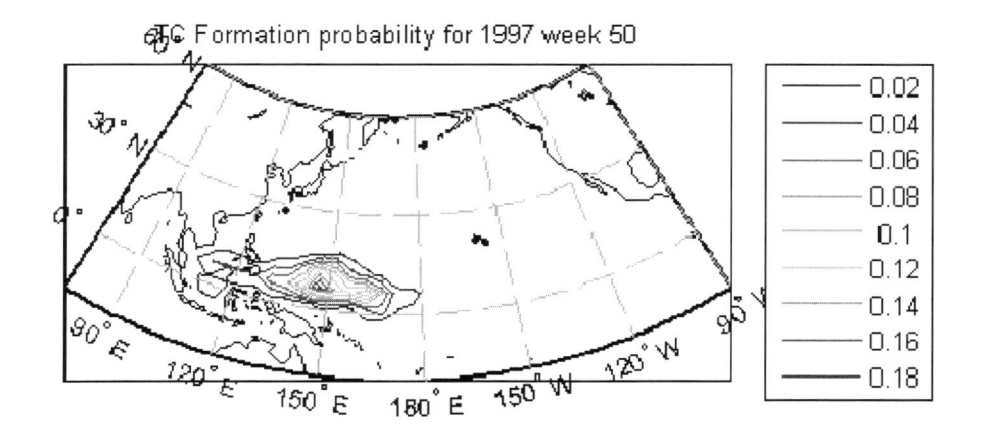

Figure 38. Model weekly probability contours for TC formation for week 50 of 1997. Red dots mark formation locations for actual TCs. Note the low probabilities compared to those in Figures 36-37. Note also that no TCs formed during this week.

While the Figures 36-38 indicate that the model is valid, the results do not constitute a true check of model validity: the model was built on data from 1980 through 2005, and thus should have good performance over those years. A more definitive check uses the estimated model to predict TC formation for years not included in the data set (similar to hindcasting in atmospheric forecasting). Even though we had excluded the 1970 through 1979 data from our model (because of differences between the data sets), it stood to reason that our model should be able to at least roughly estimate TC formation probabilities during those years. Figures 39-42 show the performance of our model in a sampling of weeks in 1974. Clearly our model works on data outside of the data set used for model development, thus substantiating the validity of the model.

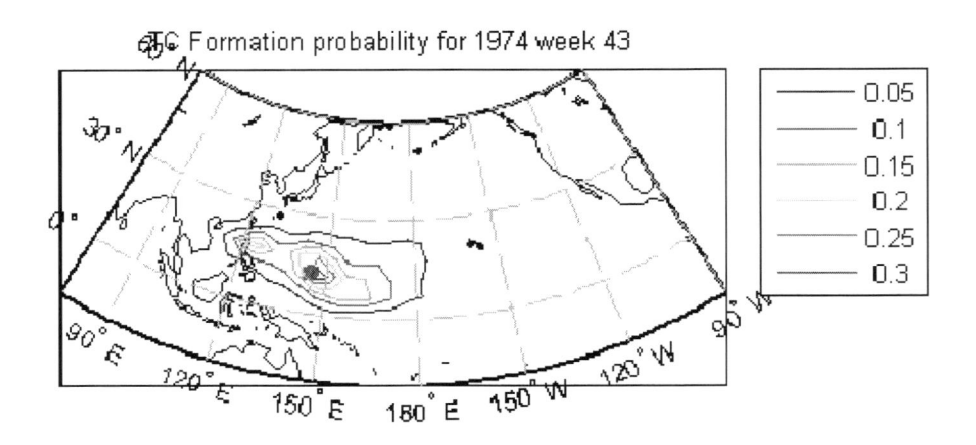

Figure 39. Model weekly probability contours for TC formation for week 43 of 1974. Red dots mark formation locations for actual TCs. TC data for 1970-1979 was not used in the model development. Thus, the results in this figure and in figures 39-42 indicate that there is a good fit between our model and actual TC formation in the Western North Pacific.

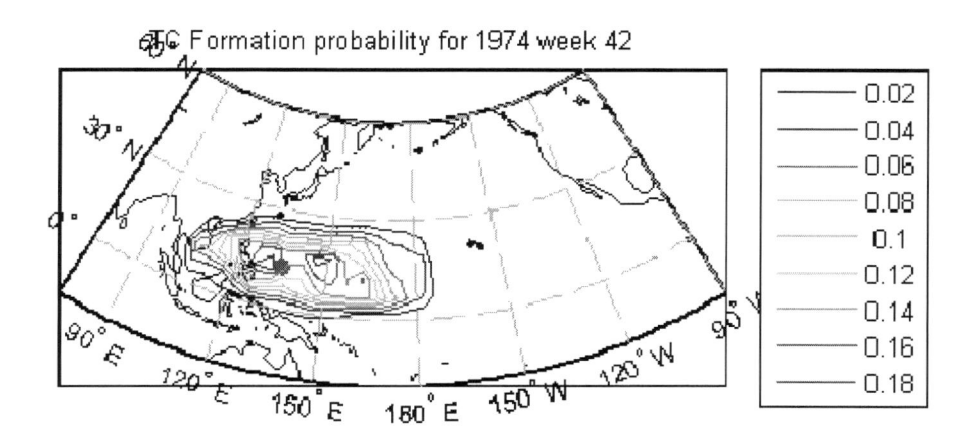

Figure 40. Model weekly probability contours for TC formation for week 42 of 1974. Red dots mark formation locations for actual TCs. TC data for 1970-1979 was not used in the model development. Thus, the results in this figure and in figures 39-42 indicate that there is a good fit between our model and actual TC formation in the Western North Pacific.

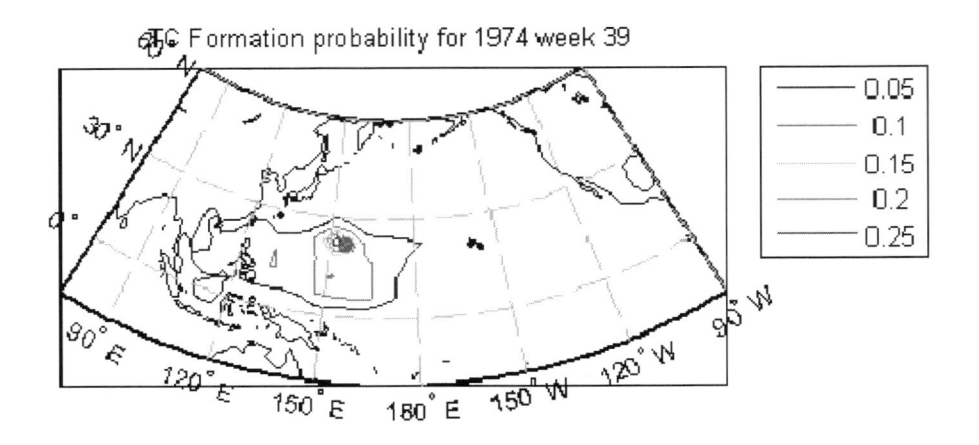

Figure 41. Model weekly probability contours for TC formation for week 39 of 1974. Red dots mark formation locations for actual TCs. TC data for 1970-1979 was not used in the model development. Thus, the results in this figure and in figures 39-42 indicate that there is a good fit between our model and actual TC formation in the Western North Pacific.

TC Formation probability for 1974 week 20

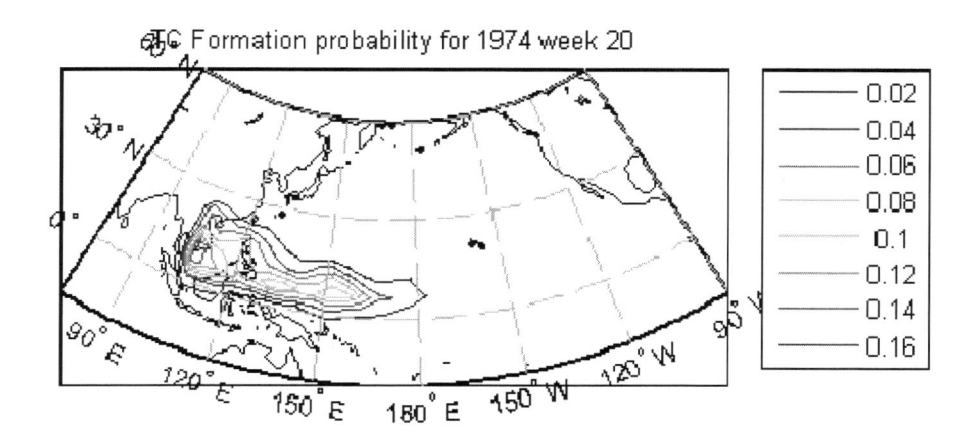

Figure 42. Model weekly probability contours for TC formation for week 20 of 1974. Red dots mark formation locations for actual TCs. TC data for 1970-1979 was not used in the model development. Thus, the results in this figure and in figures 39-42 indicate that there is a good fit between our model and actual TC formation in the Western North Pacific. Note the low probabilities compared to Figures 39-41 and the absence of any TC formations.

The probability maps in Figures 36-42 constitute some of the better matches between model probabilities and actual TC formations that we obtained. However, the overall results do demonstrate sufficient goodness of fit that we can claim that the relationships indicated by our model are valid. Appendix A and Appendix B have all the weekly probability contour plots from 2004 and 1979 respectively so that the reader can estimate for him or herself how useful a predictive tool the above model is. The model appears to have potential as a tool for predicting the number and location of TC formations (note though that the development of such a tool was not a goal of our study). However, for use as a predictive tool, the model needs further development, and more quantitative and formal assessments of its predictive skill, as discussed in Chapter IV.

g. Sensitivity of Model Probabilities to Each LSEF

Once we had a model that showed the relationship between TC formation probability and the LSEFs, we decided to investigate the sensitivity of the model probability to each LSEF, as a way of estimating the relative contribution of each LSEF to the probability of TC formation. To demonstrate this relative contribution, we selected what appear to be representative values for each of the LSEFs at the times and locations at which the model indicated high probabilities of TC formation. We then altered the LSEFs by moderate amounts (ten percent of the typical range of the LSEF over which TC formation was likely to occur) while holding the other factors constant, to see the effect on formation probability. The results are (shown in Table 4) indicate that the model's TC formation probabilities are most sensitive to changes in vertical velocity and SST, and least sensitive to changes in absolute vorticity and relative humidity.

Factor	Normal range	Base value	10% Change	Base formation probability	Probability change	Percentage change
SST (deg C*100)	2600-3000	2800	+40	0.07614	0.0208	+27.28
Abs Vort ($s^{-1}*10^5$)	2-10	6	+0.8	0.07614	0.0023	+3.05
Vert Vel (Pa/s)	0-20	5	+2	0.07614	0.0228	+29.97
RelHum (%)	60-90	70	+3	0.07614	-0.0056	-7.38
Shear (kts)	-20-20	0	+4	0.07614	-0.0102	-13.43

Table 4. Relative contribution to formation probability given a 10% change in each of the LSEFs while holding the other LSEFs constant.

h. Relationship Between Global Warming and TC Formation

As a reminder to the reader, we have shown the following:

1. Global warming appears to have a positive relationship with two of the LSEFs, SST and shear.
2. The changes in SST and shear have been in a direction considered favorable for TC formation and intensification.
3. A model for TC formation probability confirms that SST and shear are necessary independent variables for modeling TC formation probability.
4. The model indicates as SST increases or as shear tends toward zero, TC formation probability increases.
5. These results indicate that global warming has a positive relationship with TC formation probability.
6. Thus it appears that global warming induced changes in WNP SST and shear during 1970-2005 have contributed to an increase in TC formations during that period

2. ACE

To model the effect of the LSEFs on ACE, we were forced to take a slightly different approach. Examination of residual plots of linear regression fits and partial residual plots such as Figure 43 for SST, reveal that the variability of ACE increases with its expected value. Thus the modeling assumption of equal variance in multiple linear regression models is violated. Note that the ACE we analyzed occurred in a weekly five by five degree block, and it is not the total ACE for a year or the ACE for a single TC. However, just as partial residual plots for TC formation were informative, they are informative for ACE as well, as shown in the following section.

a. *Developing a Predictive Model for ACE*

Although our data does not meet the assumptions necessary to use analysis of variance, we chose to fit with model development and then check the resulting fit against an independent set of data to demonstrate the validity of the model, as we did with the model for TC formation. The data was all from weekly five by five degree blocks which had either positive values for ACE, or were zero because a TC had formed but not yet gathered sufficient strength (maximum sustained winds of 18 m/s, or 35 kts). This data set was composed of over 4500 data points. The model based on this data is shown in Table 5.

Term	Coefficient	Standard Error
Intercept	-92365.59	9113.58
SST	26.52	2.62
AbsVort	3005.25	198.86
VertVel	58.83	70.81
RelHum	211.99	59.30
Shear	55.32	42.36
Shear^2	-5.09	1.49

Table 5.   Model coefficients and standard errors for ACE.

Table 5 yields the following predictive equation for ACE:

$$\text{ACE (kt}^2\text{)} = -92365 + 26.52\text{SST} + 3005.25\text{AbsVort} + 58.83\text{VertVel} + 211.99\text{RelHum} + 55.32\text{Shear} - 5.09\text{Shear}^2 \qquad (10)$$

The ACE that is predicted by the coefficients and independent variables shown in Table 5 and Equation 10, is the ACE that is predicted to be generated in a weekly five by five degree block, *assuming a TC has formed in or is present in that block*. Again the variability of ACE increases with its expected value as shown in the partial residual plots. This forces us again to check the model fit on an independent set of data to make sure the model is valid. Thus, as we did with the TC formation data, we have excluded the data prior to 1980. The ten years of data not used in the modeling proved to be very useful for determining goodness of fit. Notice also in Table 5 that the Coriolis term is no longer part of the model as it was found to not be significant. This is not surprising, since this model describes ACE given that a TC (which is dependant on Coriolis acceleration) has already formed.

We expected that the same independent variables that would be important for TC formation would be important for the generation of ACE. Certainly Table 5 bears this out, since the independent variables that are important for formation are also important for ACE (except for the Coriolis term). The major difference is that the strong, positive relationship between ACE and relative humidity is much more straightforward than the relationship we observed for TC formation. Also, exactly as expected, and consistent with our TC formation model, increasing SST, humidity, vorticity, and vertical velocity (up to our observed threshold limits) all tend to increase ACE, while any increase in the magnitude of shear tends to reduce ACE.

We then proceeded to check our model that predicts ACE for a block given in which a TC had formed and was present, using our LSEF data as input for the model. We compared the annual actual ACE for each year to the annual predicted ACE for each year. The results are shown in Figure 43.

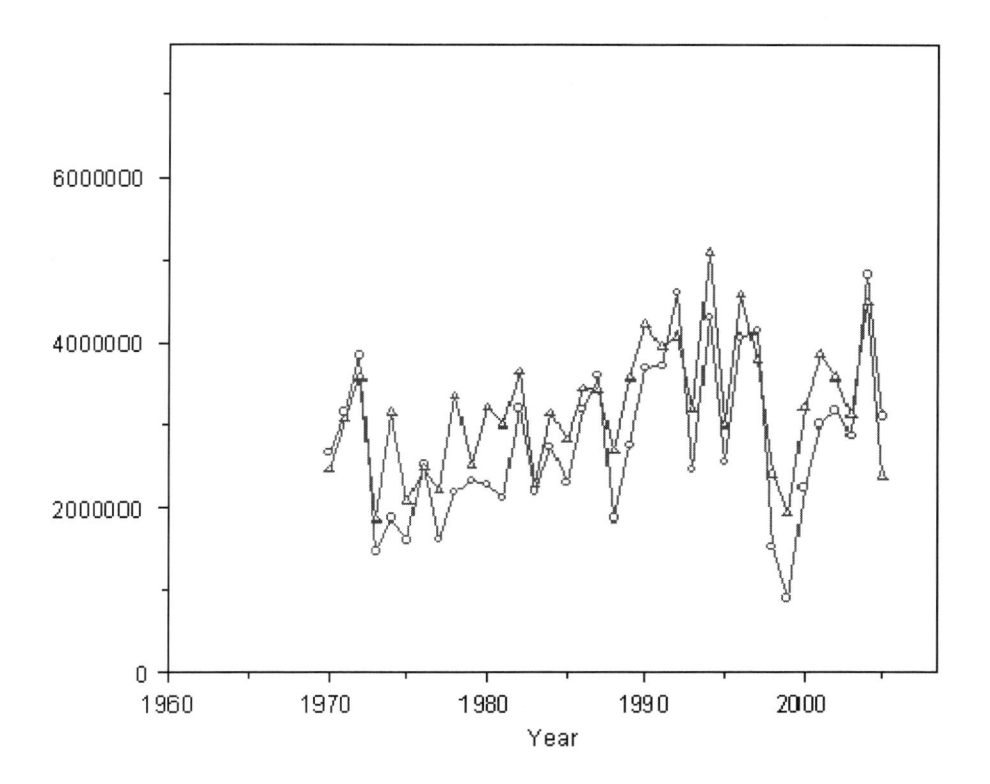

Figure 43. Annual ACE values for 1970-2005. Actual (in blue) and modeled (in red).

Note that included in Figure 43 are actual (in blue) and predicted (in red) values for ACE for the years 1970 through 1979. The data from these years was not included in the model fit. Moreover, as noted earlier in this chapter, there are some rather significant differences between the Reanalysis I and Reanalysis II data sets. Thus, it is encouraging that the predicted values of ACE match actual ACE rather closely during 1970-1979.

Finally, Figure 43 seems to validate our model for ACE. There are clearly some differences between the model and actual values of ACE, but for the most part, both the magnitude and the trend (increasing or decreasing from year to year) of the model seems to be accurate enough to state with confidence that the model works and that the relationships between the LSEFs and ACE hold true. Of course, additional quantitative assessment of the model skill would be useful.

b. *ACE Partial Residual Plots - SST*

Figure 44 shows the partial residual plot for ACE against SST. The straight horizontal line centered on zero represents all the weekly five by five degree blocks for which there was no TC activity (on the order of 140,000 data points). All data above the horizontal line represents the value of ACE in all the weekly five by five degree blocks for which there was TC activity at its corresponding SST. ACE will be zero if a TC did not form (in some cases ACE is zero the week a TC formed if it did not gather enough strength to reach 35 kts prior to week's end).

Note that the ACE residual plot for SST has a different pattern than the corresponding figure for TC formation (compare Figures 44 and 31). In particular, the non-zero ACE cases occur in SST ranges that are different from, but similar to, those for TC formations. This is because there is not a one for one correspondence between the weekly five by five blocks for TC formation and for ACE, since formations occur when the TC is below the intensity threshold for the calculation of ACE (maximum sustained winds of 18 m/s or 35 kts; see Chapter II). Also, the non-zero ACE cases are spread over a wide range of the vertical axis. That is, the magnitude of ACE is a function of SST (as shown by Figure 44). So not only do higher SSTs make it more likely there will be a TC, they make it more likely that the strength of the TC will be greater too.

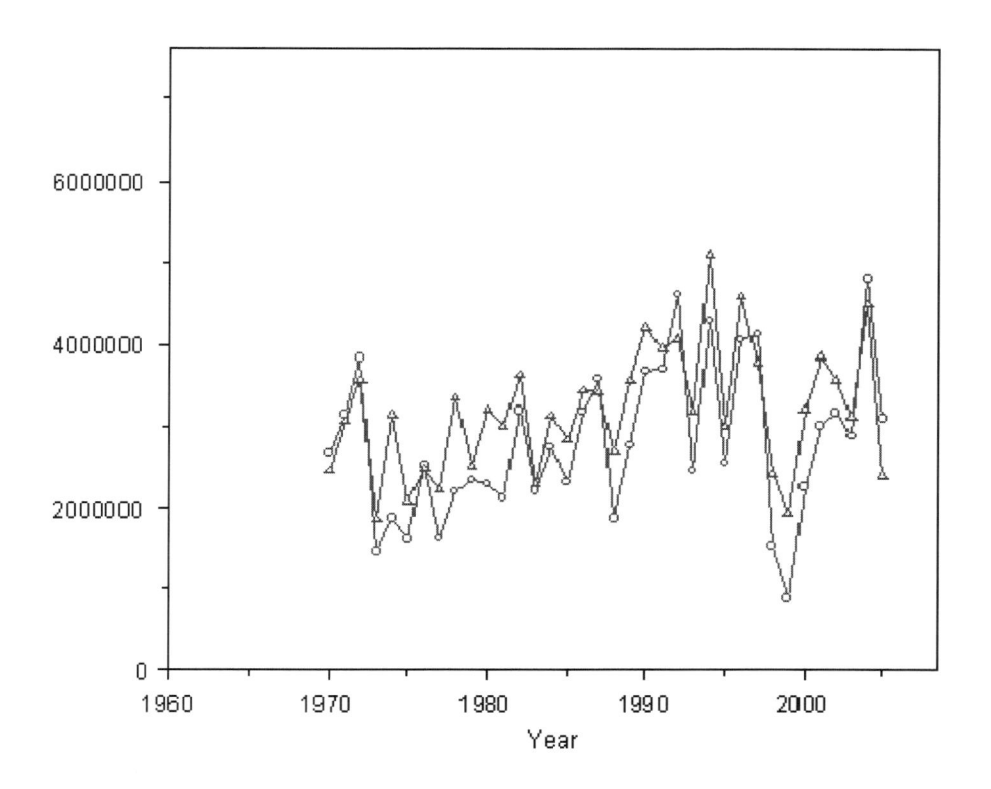

Figure 44.   ACE partial residual plot for SST.

c. Shear Partial Residual Plot for ACE

For shear and the remainder of the LSEFs the magnitude of ACE can be quite large over a relatively broad range of LSEF values, so long as these values are at least near the thresholds identified in the TC formation residual analyses (see Figures 45 through 48). For example, as shown in Figure 45, as long as the value for shear is between -20 and 20, the value of ACE can be quite large or it can be quite small (in fact it is skewed toward small). But large values of ACE can and do occur anywhere within that range of shear. As we saw with TC formation, ACE peaks near a shear magnitude of zero.

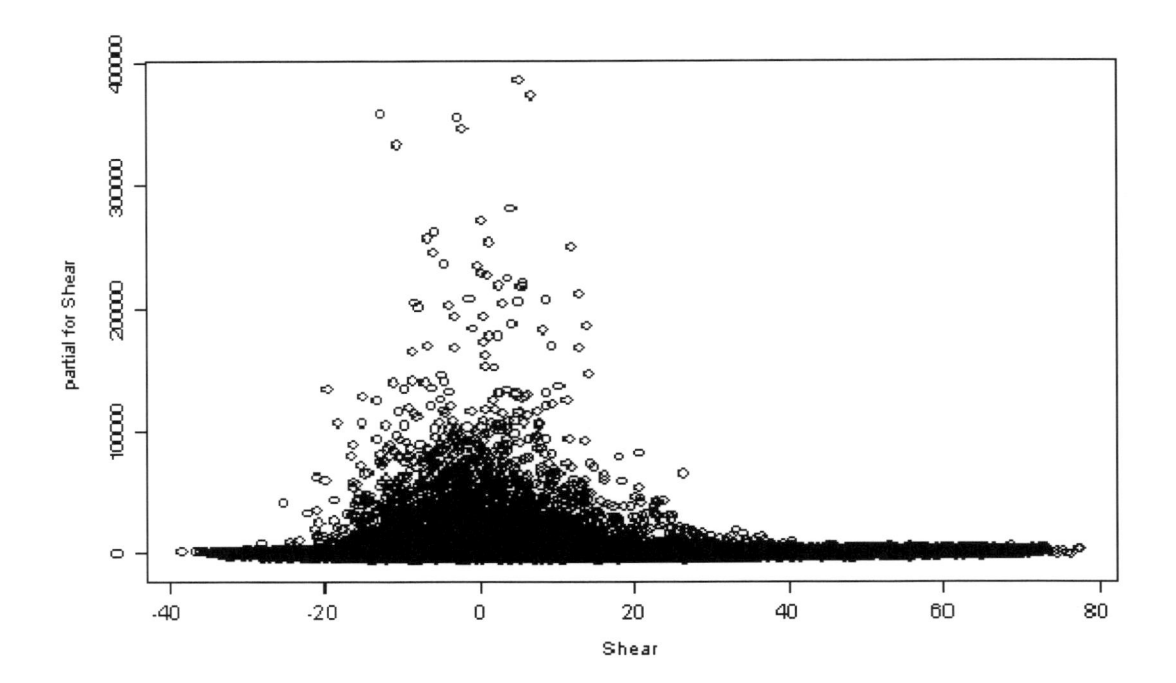

Figure 45.   ACE partial residual plot for shear.

d. Relative Humidity Partial Residual Plot for ACE

Similar to the results for TC formation, a relative humidity of about 55-90% appears to be favorable for the generation of ACE (see Figure 46).

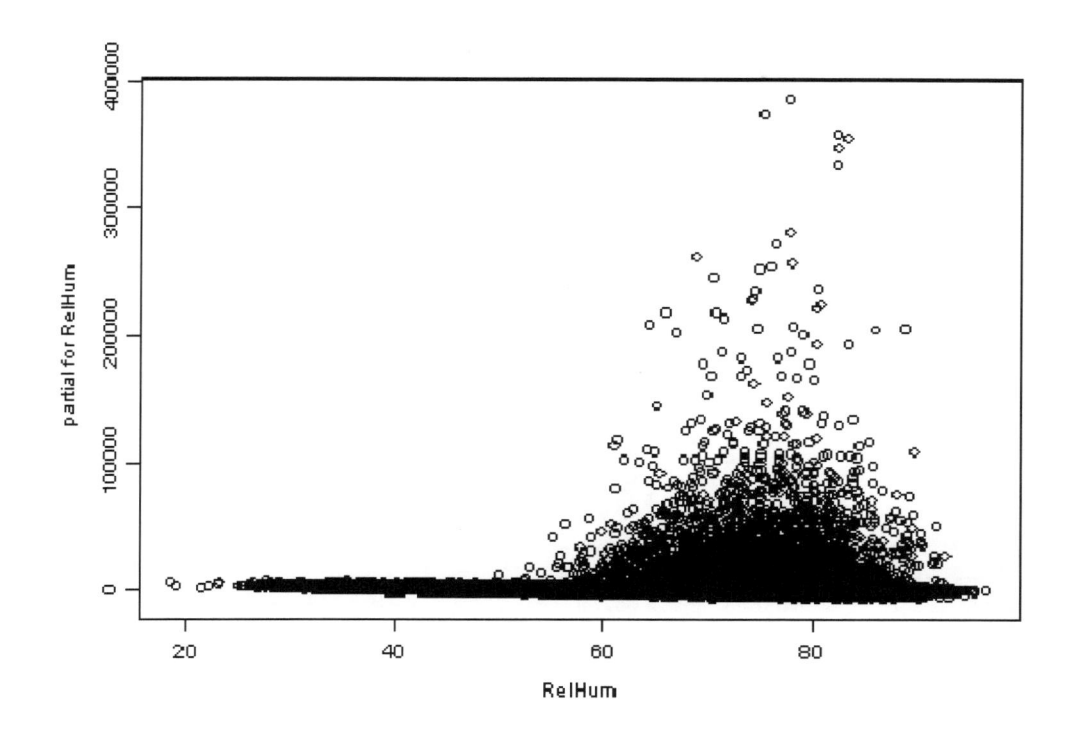

Figure 46. ACE partial residual plot for relative humidity.

e. Vertical Velocity Partial Residual Plot for ACE

Figure 47 shows that a zero or positive value of vertical velocity is normally necessary to generate ACE. It is intriguing again that, as with TC formation, it seems that there can be too much vertical velocity, as the ACE residuals taper off sharply for vertical velocities greater than 22 Pa/s.

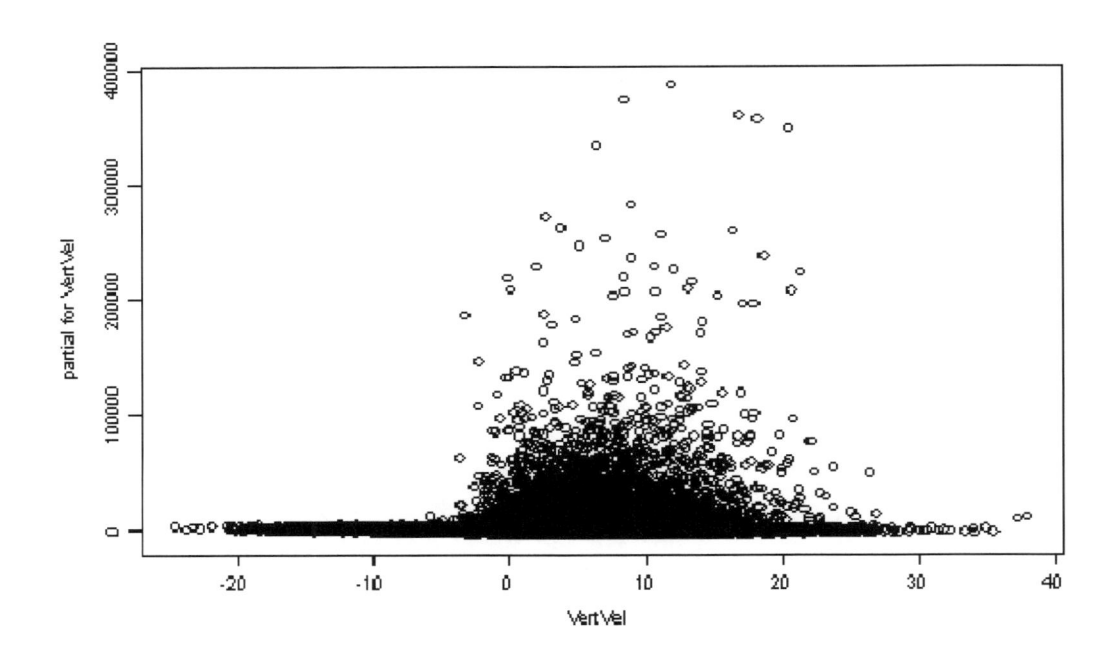

Figure 47. ACE partial residual plot for vertical velocity.

f. Absolute Vorticity Partial Residual Plot for ACE

As we saw with TC formation, the peak value for absolute vorticity for which most ACE occurs at is about 5-6 $s^{-1}*10^5$ (see Figure 48). ACE probably decreases beyond this peak as explained earlier: a high absolute vorticity most likely means the data is from a higher latitude, where there are lower SSTs and higher shear magnitudes, both unfavorable to TC formation and intensification.

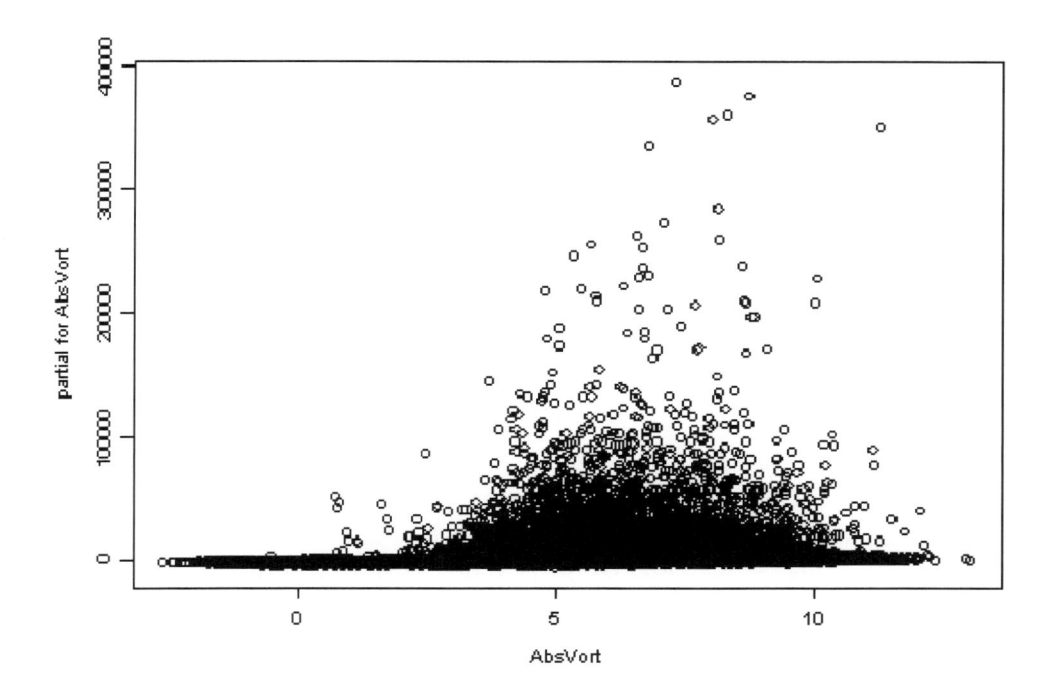

Figure 48. ACE partial residual plot for absolute vorticity.

g. Sensitivity of Model ACE to the LSEFs

We investigated the sensitivity of the model predicted ACE to the LSEFs, as a way of estimating the relative contribution of each LSEF to the generation of ACE in a weekly five by five degree block. We selected the same representative values for each of the LSEFs as we did for TC formation (the range and peaks for the LSEFs are consistent for both TC formation and ACE), and altered them by reasonable amounts (again, 10%) to see the effect on ACE generation. The results are shown in Table 6, with the two largest contributors being (in order) absolute vorticity and SST. Note that shear is a very slight positive contributor to ACE if the magnitude of shear is near zero. If the magnitude of shear grows significantly from zero, shear becomes an increasingly negative contributor to ACE.

Factor	Normal range	Base value	Change	Base ACE (kts^2)	ACE change (kts^2)	Percentage change
SST (deg C*100)	2600-3000	2800	+40	14999	1060	+7.10
Abs Vort ($s^{-1}*10^5$)	2-10	6	+0.8	14999	2404	+16.03
Vert Vel (Pa/s)	0-20	5	+2	14999	117	+0.78
RelHum (%)	60-90	70	+3	14999	636	+4.20
Shear (kts)	-20-20	0	+4	14999	139	+0.93

Table 6. Relative contribution to ACE given a 10% change in each of the LSEFs while holding the other LSEFs constant.

h. *Relationship Between Global Warming and ACE*

In summary, we have shown the following:

1. Global warming appears to have a positive relationship with two of the LSEFs, SST and shear.
2. The changes in SST and shear are in a direction favorable for the generation of ACE.
3. A model of ACE based on the LSEFs confirms that SST and shear are necessary independent variables for modeling ACE.
4. The model indicates as SST increases or as shear tends toward zero, ACE tends to increase.
5. Therefore, it appears that global warming has a positive relationship with ACE.
6. Thus, it appears that global warming induced changes in WNP SST and shear during 1970-2005 have contributed to an increase in ACE during that period.

IV. CONCLUSIONS AND RECOMMENDATIONS

A. TRENDS IN THE LARGE SCALE ENVIRONMENTAL FACTORS

We showed in Chapter III that WNP SSTs have increased during 1970-2006 (Equation 3). We also saw that WNP shear has increased, but that the absolute value of shear has actually decreased, particularly in the tropics (Equation 4). These assessments are on a large scale (approximately) ten by forty degrees, and on a seasonal timeframe. But if the LSEFs really do influence TC activity, then two of the LSEFs have changed in directions that would tend to increase TC activity, while the others two have been essentially unchanged. Therefore, an increase in TC activity would be expected in the data set. In addition, one might expect as greenhouse gasses continue to accumulate in the atmosphere that the observed trends in the LSEFs would continue as well.

For SST and shear, we observed that the changes occurring over 1970-2006 were consistent with, and apparently attributable to, global warming. However, these changes did not occur uniformly across the entire WNP (Figures 12 and 14). Instead, they changed by different magnitudes and in different directions depending on the region of the WNP. Thus, analyzing the LSEFs by averaging the LSEFs over large geographic regions or periods of time, as done in prior studies of the impacts of global warming on TC activity, smoothes out these sub-basin scale variations and potentially eliminates the ability to detect important effects of the LSEFs on TC activity.

B. TRENDS IN TROPICAL CYCLONE ACTIVITY

The observed long term changes in the LSEFs suggests that TC formations and ACE should have also increased during 1970-2005. Indeed, that is the case as we see an increase in actual TC formations, and an increase in ACE as well (Figures 23 and 25). It is important to realize that there is an increase in ACE that is independent of that due merely to the fact that the number of TCs is increasing. A key point is that the average number of TCs per year has increased during 1970-2005, and that ACE per TC on average has increased as well (Figure 27).

We have also identified a series of relationships between global warming measures, the LSEFs, and TC activity measures. We have seen a strong linear relationship between atmospheric greenhouse gasses and GASTA, one of our measures of global warming (Figure 3). We have also determined that there is a relationship between global warming and the LSEFs (Figures 10, 13, 15, 17 and 22): SST is increasing, and shear is increasing (corresponding to a decrease in the magnitude of shear in the tropics). The remainder of the LSEFs have remained essentially unchanged. We have confirmed that the LSEFs (SST, shear, vertical velocity, relative humidity, and absolute vorticity) do influence TC activity, both formation and ACE (see Chapter III sections C.1 and C.2). This suggests that if Earth continues to warm, tropical cyclones in the WNP will become on average more frequent, and on average more powerful. However, the patterns of SST and shear changes in the WNP indicate that future increases in shear may eventually cause tropical regions of weak negative shear to become regions of positive shear, thus potentially offsetting the effects of increasing SST.

It has been shown that TC activity is indeed a function of all the LSEFs. To attempt to characterize TC activity with only one or two of the factors is like assessing the value of a house based only on its number of bathrooms. Likewise, assessing the impacts of global warming on TC activity using only one of the LSEFs (e.g., SST, as in almost all prior studies of the impacts of global warming) is likely to produce misleading results.

It has been shown that as a factor, SST is unique in that it is the only factor for which an increase raises formation probability and ACE values without apparent bound (Figures 31 and 44). All other LSEFs appear to have limits beyond which TC activity begins to decrease. For example, there are values of absolute vorticity and vertical velocity beyond which TC activity tends to decrease (Figures 34, 35, 47, 48). Non-zero shear tends to lower TC formation probability and ACE (Figures 32 and 45).

We determined that for a same percentage change in any of the LSEFs, the factor with the greatest effect on TC formations is vertical velocity, followed by SST (Table 4). Absolute vorticity comes in third though at only about 10% of the contribution that SST and vertical velocity make. Relative humidity has a slight negative contribution to formation, and as expected, any shear whatsoever reduces formation probability.

We also determined that for a same percentage change in any of the LSEFs, the factor with the greatest effect on

ACE is absolute vorticity, followed by SST (Table 6). Relative humidity, a positive contributor to ACE comes in third. In addition, ACE is not nearly as sensitive to non-zero shear as TC formation is.

The common belief is that the LSEFs are necessary but not sufficient to cause or sustain TC activity. However, the probability contour plots matched with actual TC formations, hints that the LSEFs, in particular, high SSTs, may be both necessary *and* sufficient for TC formation, at least in the WNP. This idea of course conflicts with the well established concept that TC formation requires some combination of both the necessary LSEFs, and a TC triggering process (e.g., an easterly wave, Elsberry 1995). To test the idea that the LSEFs may be both necessary and sufficient, more rigorous quantitative assessments of the TC formation model's skill need to be conducted.

The results of Chapter III, sections C.1 and C.2, show that the LSEF conditions favorable for TC formation and ACE in the WNP should be slightly altered for some of the LSEFs, leading to the following:

- SST: 26 °C or greater (unchanged).
- Vertical wind shear of the horizontal winds: weak, where "weak" is defined as a magnitude of 20 m/s or less.
- Absolute vorticity: zero or positive but less than 13 $s^{-1}*10^5$ (in the northern hemisphere).
- Vertical velocity: zero or upward but not exceeding -30 Pa/s.
- Relative humidity at mid levels: in excess of 55%.

Note that none of these changes are radical departures from the LSEFs defined early in this study. However, our results allow us to quantify the LSEFs more carefully based on more extensive data sets and analysis methods than in prior studies. For example, our results allow such descriptions of TC activity favorable conditions as "large levels of absolute vorticity" and "weak vertical shear" to be more carefully quantified and to be bounded by observationally based thresholds.

We found that predicting formation probabilities and ACE values was very sensitive to changes in the generation of the reanalysis fields, and as such, Reanalysis I data needed to be excluded from the model formulation, though it was of sufficient quality to be used in quality of fit tests following model development. Extending Reanalysis II data back in time to at least 1970 would be very helpful for future research.

The TC formation probability contour plots show that TC formation is a function of "right here, right now" LSEF data. That is, to attempt to characterize TC formation with seasonal and/or basin wide averaged values significantly dilutes the contribution each LSEF makes towards TC formation probability. And, since two of the LSEFs appear to be affected by global warming, it is important to account for at least weekly and five by five degree variations in the LSEFs when trying to assess the impacts of global warming on TC activity in the WNP.

It was never our intention to develop a model so accurate that it could be used for forecasting the number and locations of TC formations. However, the probability contour plots (Figures 36 through 42) raise the possibility that statistically based forecasting of TC formation of the sort used in this study may indeed be possible, and with reasonable accuracy.

C. LESSONS LEARNED

Assumptions were made regarding what the measures of TC activity should be, and as many authors in the literature search used ACE, we used ACE as well for the sake of consistency. Using wind speed squared when wind speed is 35 kts or greater makes ACE a rather nonlinear measure of the strength of a TC. In addition to the previously noted differences between how wind speed is measured by organization, calculating ACE without the storm radius (a measure normally missing from the best track data set) makes ACE a less than ideal measure of TC intensity.

In hindsight, both ACE and PDI[48] (power dissipation index - a measure similar to ACE that uses the cube of wind velocity instead of the square) may not be the best of measures for what humanity cares about. ACE and PDI may quantify the intensity of a TC over its life, but what humans generally care more about is the power of a storm at landfall and for the next 12 to 72 hours. If a strong storm churns away at sea for weeks at a time but never makes landfall, ACE for the season will rise dramatically but with the exception of the hazards to aviators and mariners at sea, the impact of that TC on humanity will be negligible.

Keep in mind also that TC damage on landfall (monetarily) is determined largely by population density, wealth, and planning[49], and not by the size or intensity of

---

[48] K. Emmanuel, *Nature,* **436**, 686-688 (2005).

[49] R. A. Pielke Jr, C. Landsea, M. Mayfield, J. Laver, R. Pasch, *Bull.Amer.Meteor.Soc* **86**, 1571 (2005).

the TC. One possible improvement on ACE and PDI is integrated kinetic energy (IKE) developed by Powell and Reinhold[50].

We think that the models developed in our study provided more accurate assessments of the impacts of global warming on TC activity than those of prior studies. This is in large part because we did not use basin wide and seasonal average values for the LSEFs and TC activity, but instead used weekly five by five blocks of data. However, we used monthly mean SSTs for our raw SST data (for the reasons given in Chapter II). This monthly mean constraint may have restricted what our models could do, especially since SST appears to be a very important LSEF. In future studies, it would be worth the effort to interpolate the monthly data into weekly averages. Another option available is the NCEP OISST data set (six hourly and 1 degree resolution). We had rejected this data set as an option because it only extended back to 1981 (Table 1). But since our models ended up being based on Reanalysis II data back to 1980, this limit at 1981 now does not seem as severe as when we began exploring data sets.

This research would have been far easier had the lead author taken courses in data mining and non-parametric statistics. Should others pursue a follow on topic from this one, these courses would be a worthwhile investment of their time.

D. RECOMMENDATIONS FOR FUTURE STUDY

This work just barely begins to scratch the surface of the wealth of information that can be gleaned from our data

---

[50] M. D. Powell, T. A. Reinhold, *Bull.Amer.Meteor.Soc* **88**, 513 (2007).

set. Subjects for future study should include, but not be limited to, the following questions and topics:

1. Can similar modeling be done in other regions such as the Atlantic? Would the same results be generated?

2. How similar are the TC formation and ACE generation processes, as revealed by the regression models, for the different tropical basins?

3. Can similar modeling be done using the JMA and other TC data sets? Would the same results be generated?

4. Use regression to show a relationship between JTWC data and JMA data (as well as other TC data bases). The differences between existing TC data sets are adding a lot of confusion as to what the long term patterns are in TC activity.

5. We have seen that temporal changes in the LSEFs are region dependant. Can we quantify those changes? Can we predict those changes for other parts of the planet?

6. Refine the models developed in this work. If at all possible, use a quality SST data set at a higher temporal resolution.

7. Is it possible that the TC formation model or the ACE model could be used as an aid for medium and long range forecasting of TC formations and intensity?

APPENDIX A. TC FORMATION PROBABILITY CONTOUR PLOTS FOR WEEKS 20-52 OF 2004

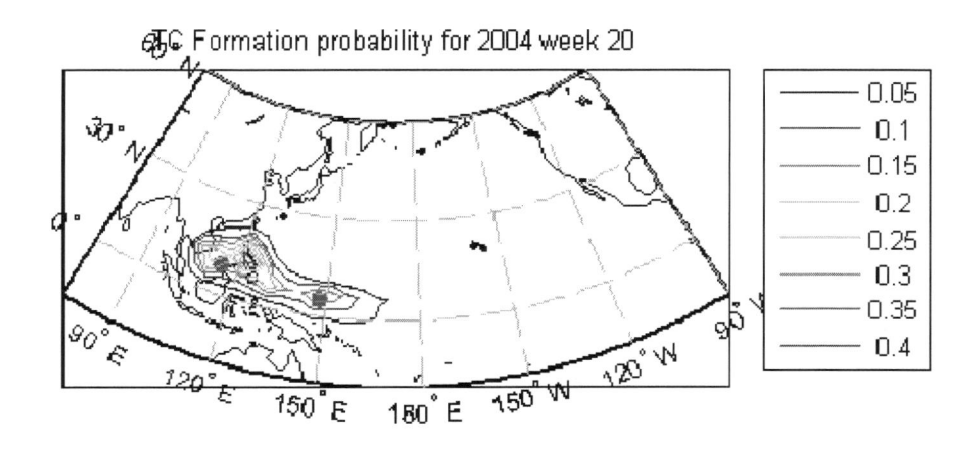

TC Formation probability for 2004 week 20

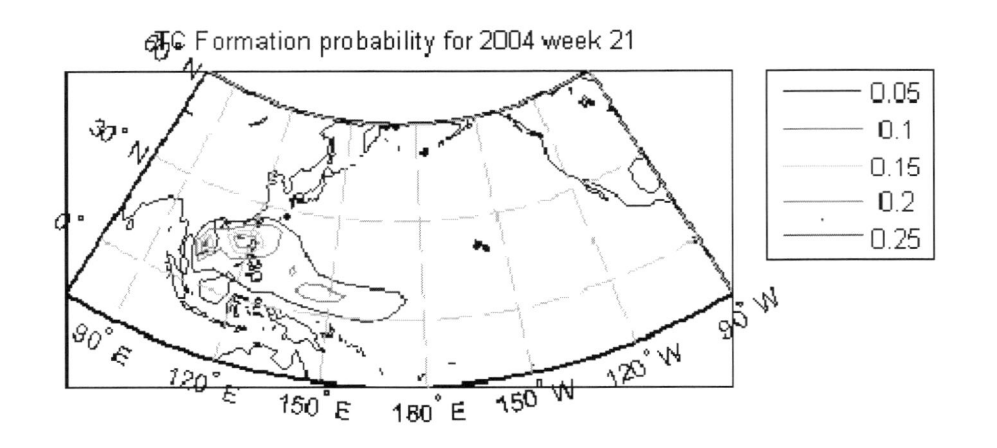

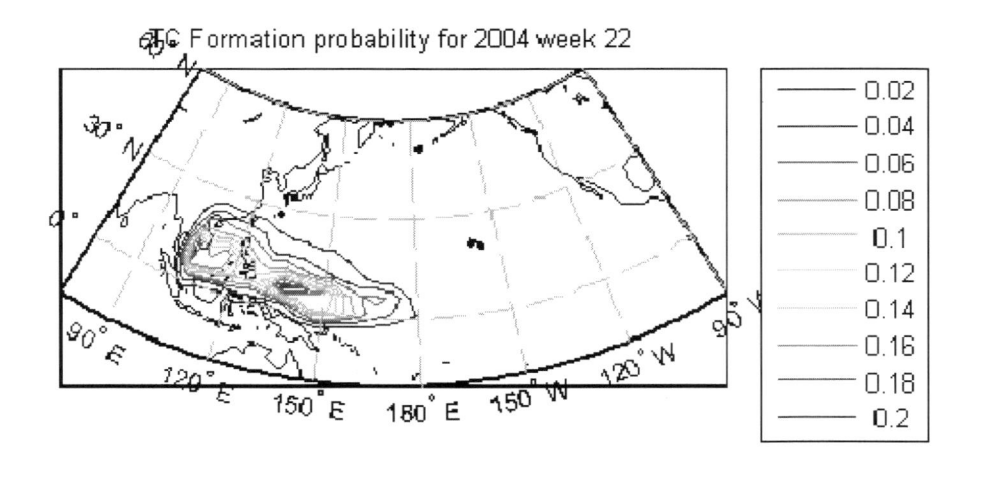

TC Formation probability for 2004 week 21

TC Formation probability for 2004 week 22

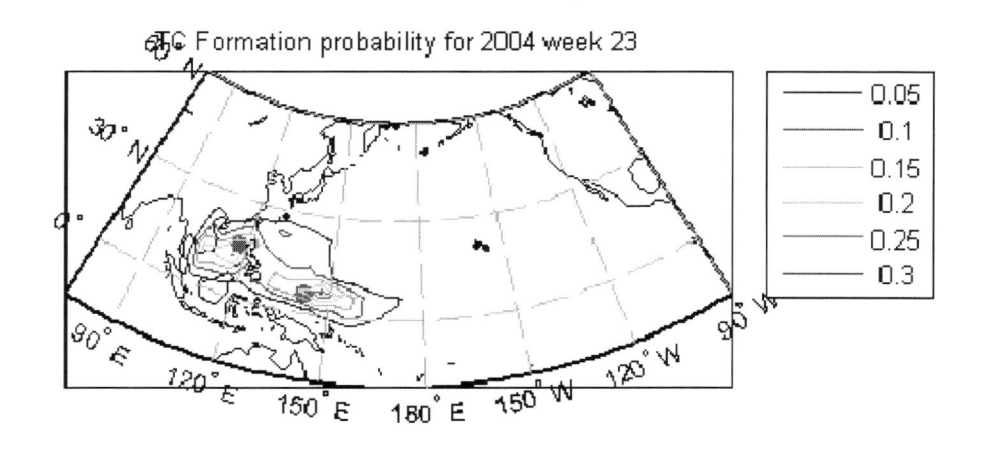

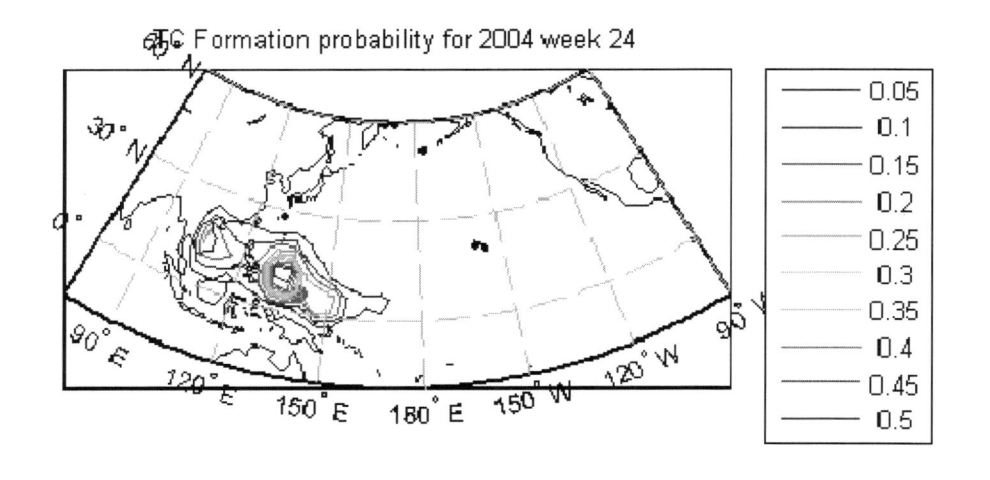

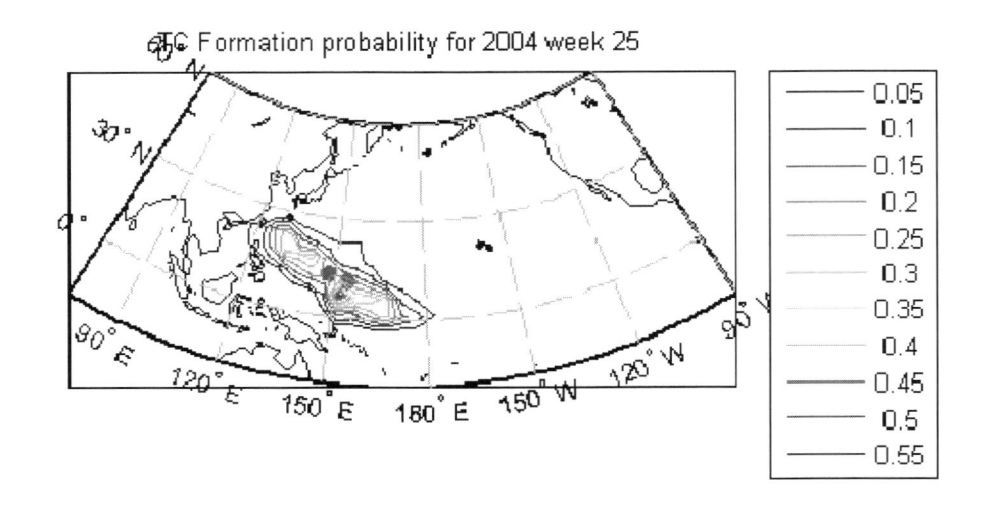

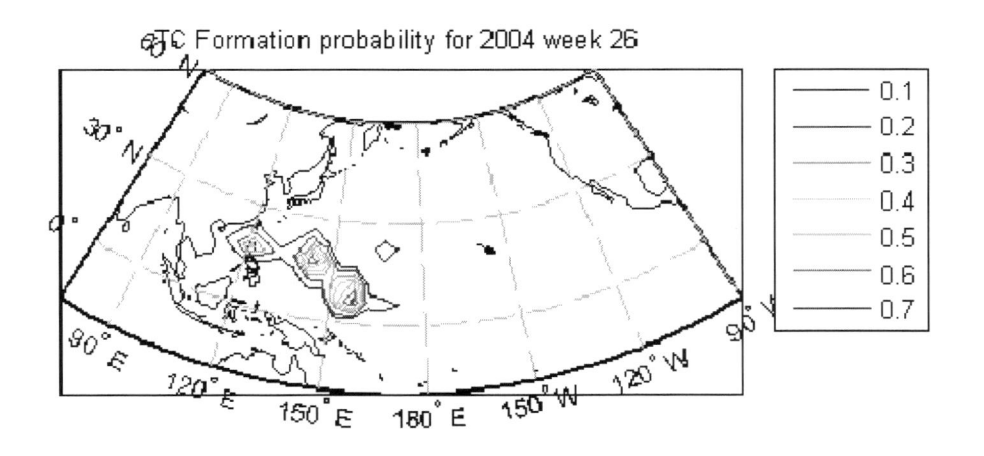

TC Formation probability for 2004 week 25

TC Formation probability for 2004 week 26

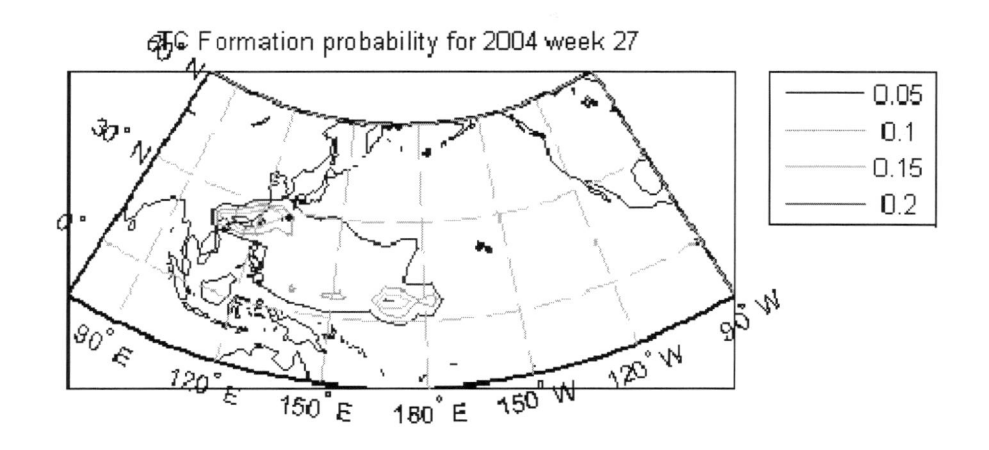

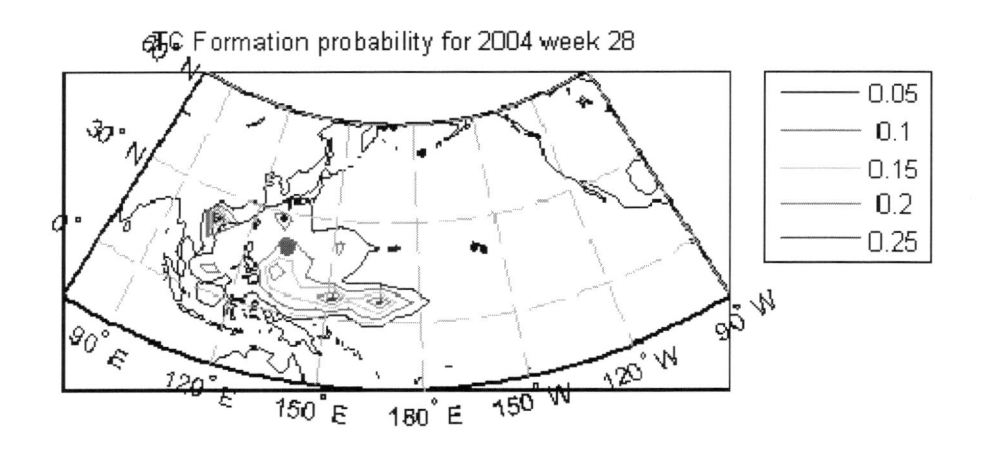

TC Formation probability for 2004 week 27

——	0.05
······	0.1
——	0.15
——	0.2

TC Formation probability for 2004 week 28

——	0.05
——	0.1
——	0.15
——	0.2
——	0.25

TC Formation probability for 2004 week 29

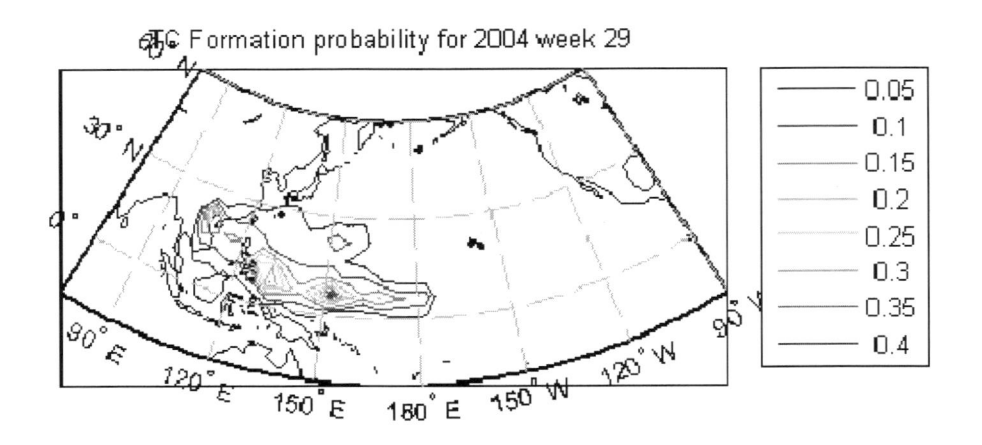

TC Formation probability for 2004 week 30

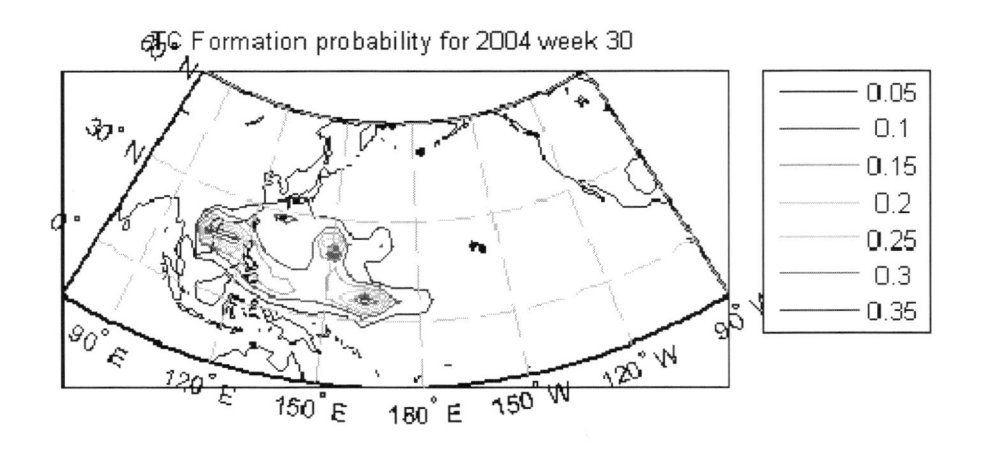

TC Formation probability for 2004 week 31

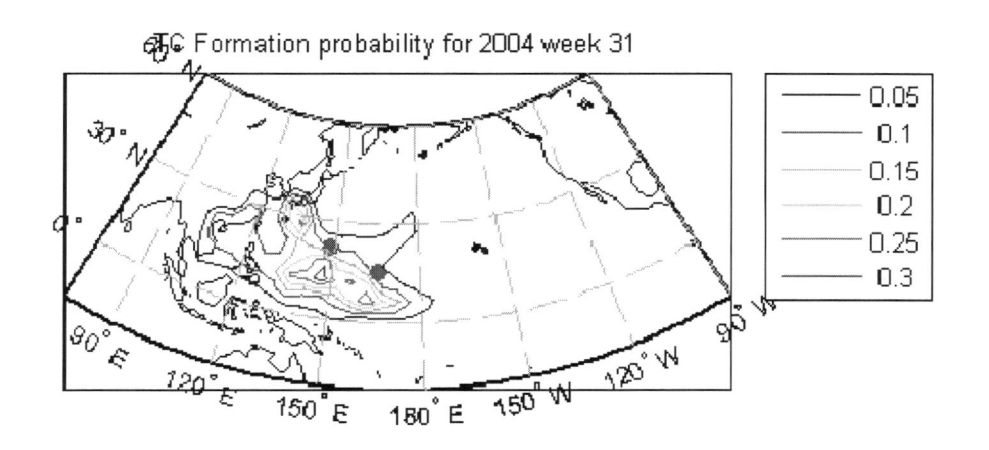

TC Formation probability for 2004 week 32

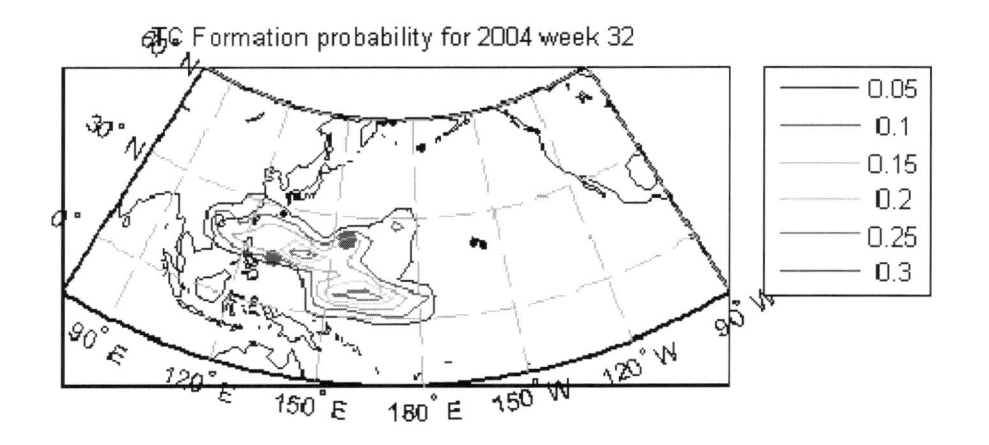

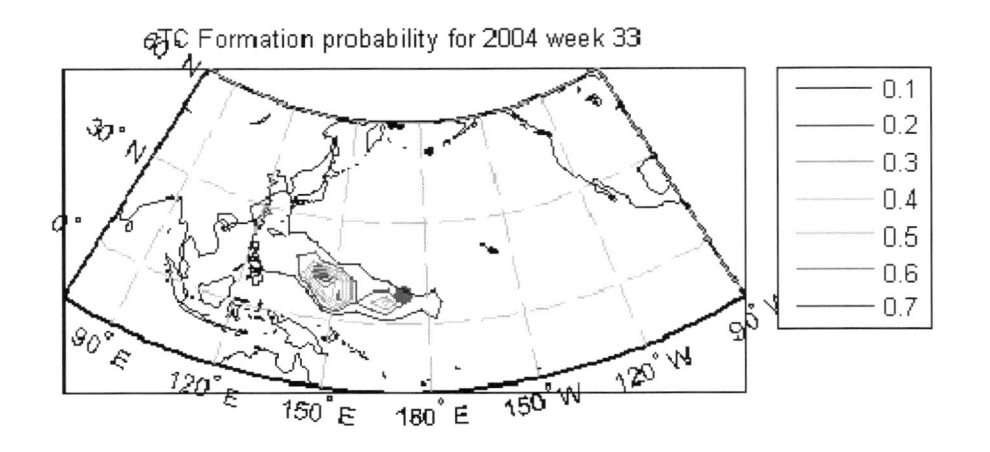

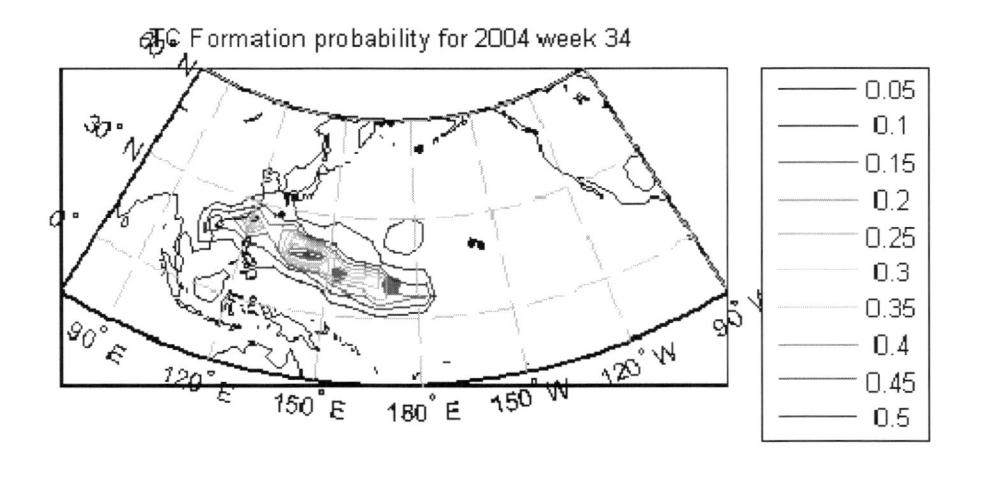

TC Formation probability for 2004 week 33

TC Formation probability for 2004 week 34

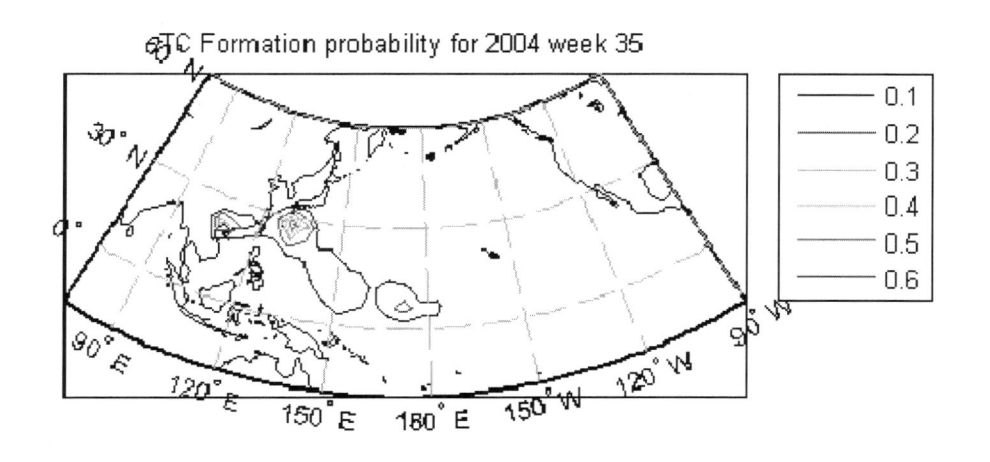

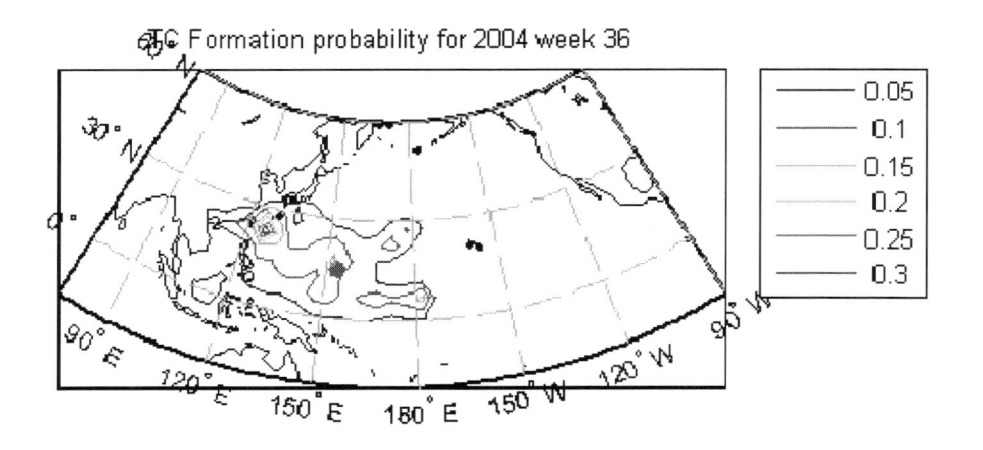

TC Formation probability for 2004 week 35

TC Formation probability for 2004 week 36

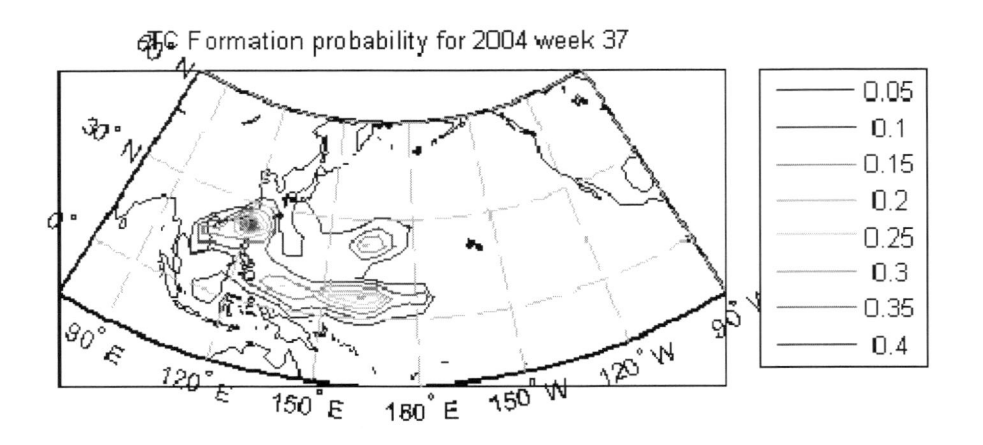

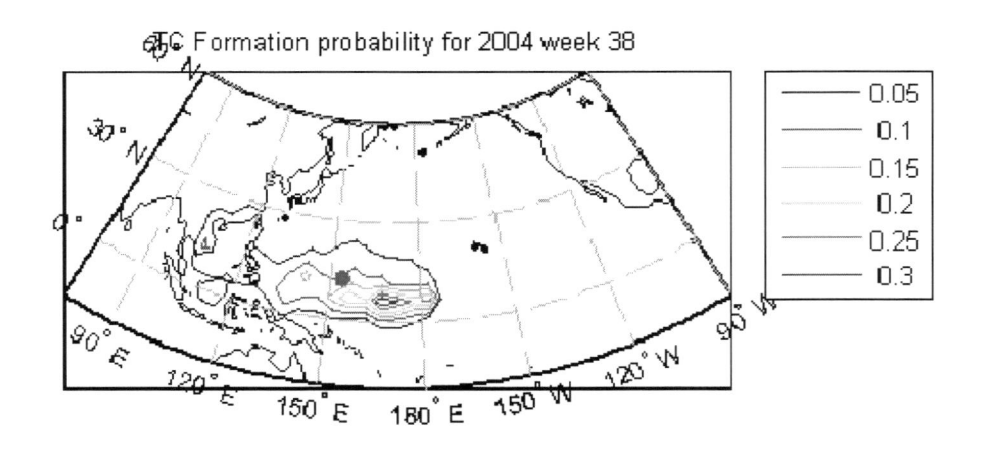

TC Formation probability for 2004 week 37

TC Formation probability for 2004 week 38

TC Formation probability for 2004 week 39

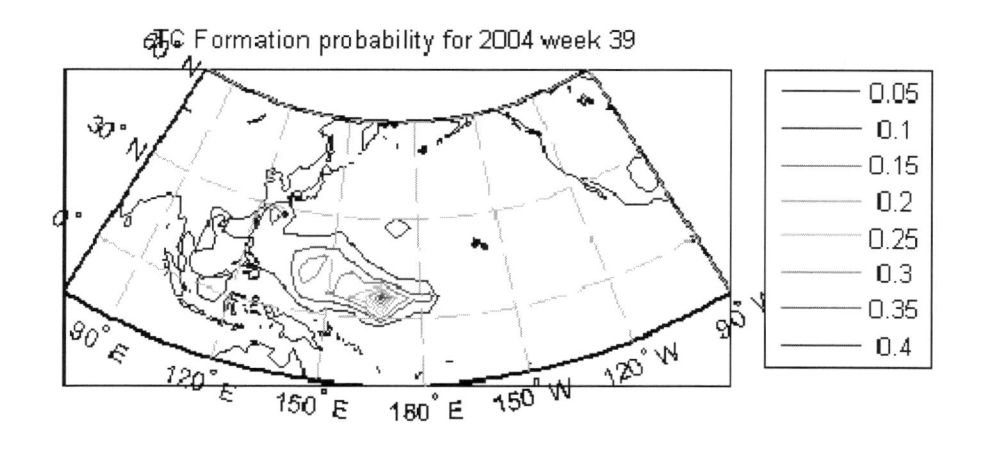

TC Formation probability for 2004 week 40

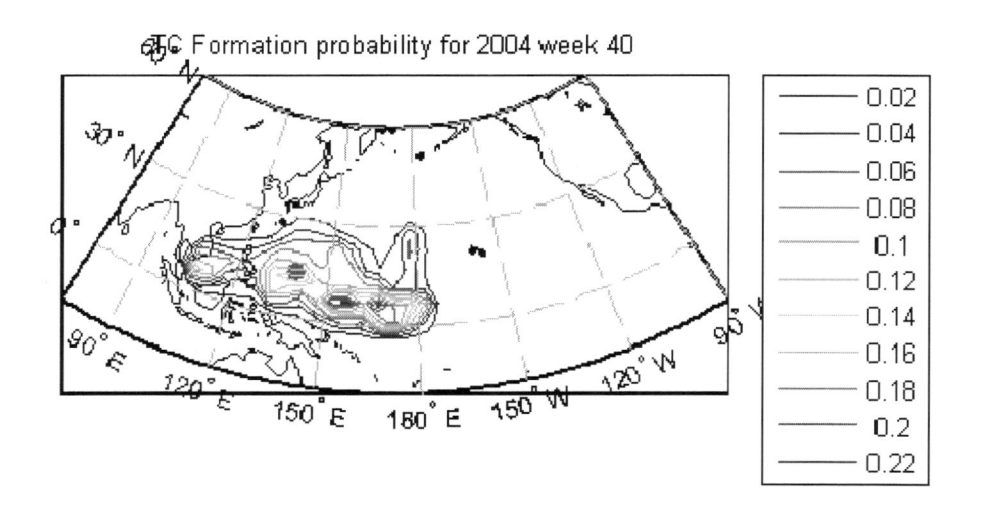

TC Formation probability for 2004 week 41

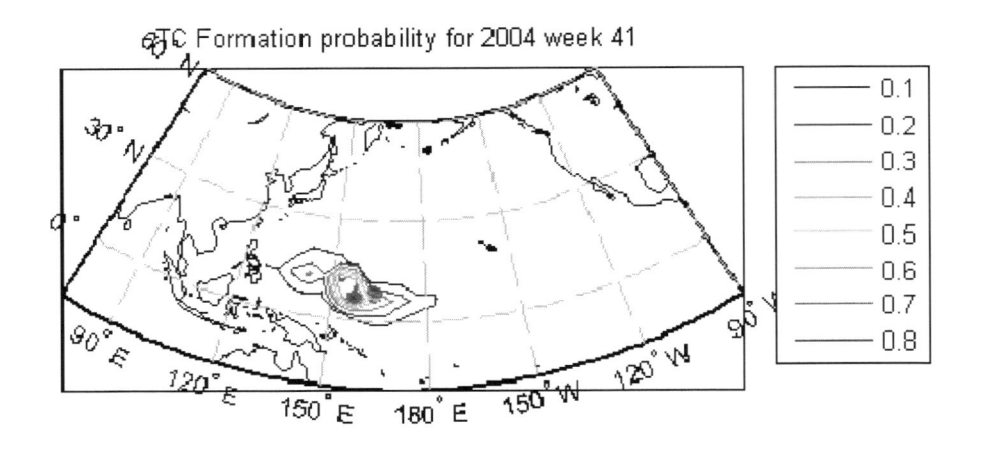

TC Formation probability for 2004 week 42

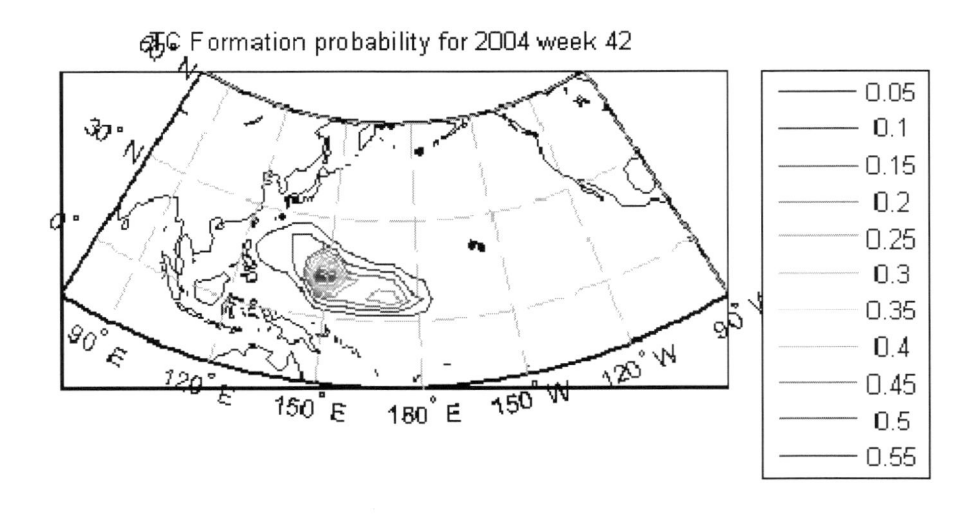

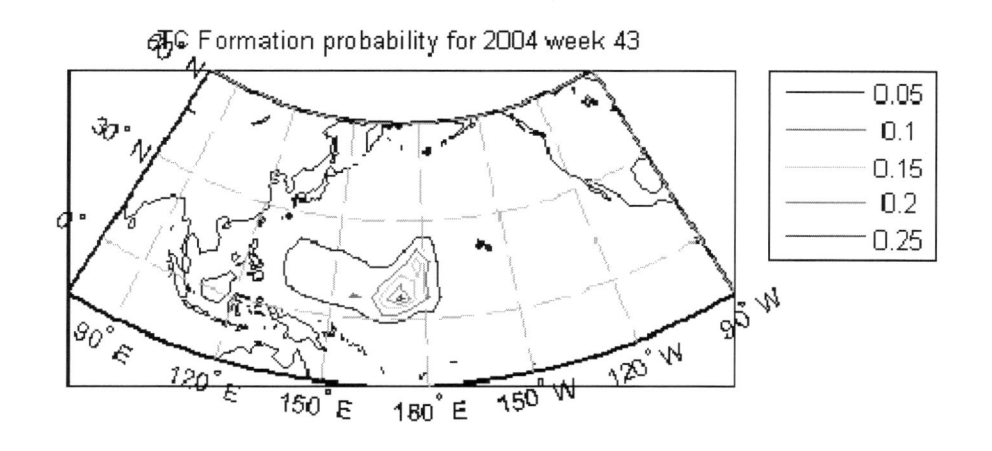

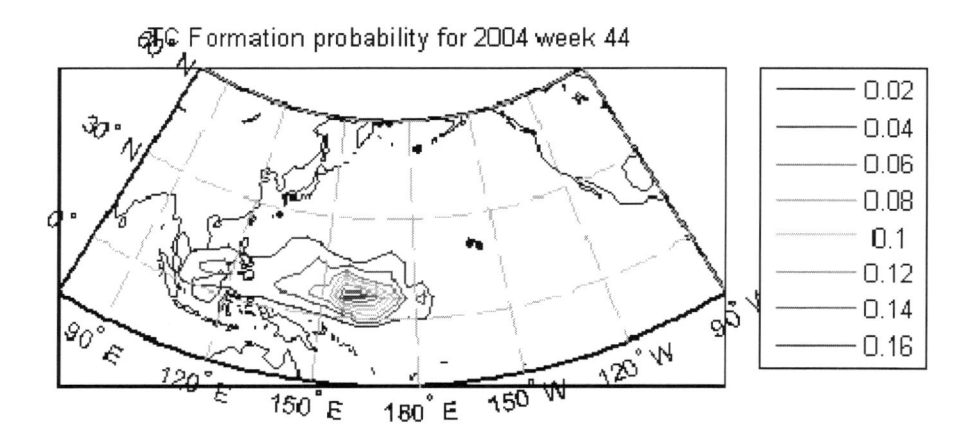

TC Formation probability for 2004 week 43

TC Formation probability for 2004 week 44

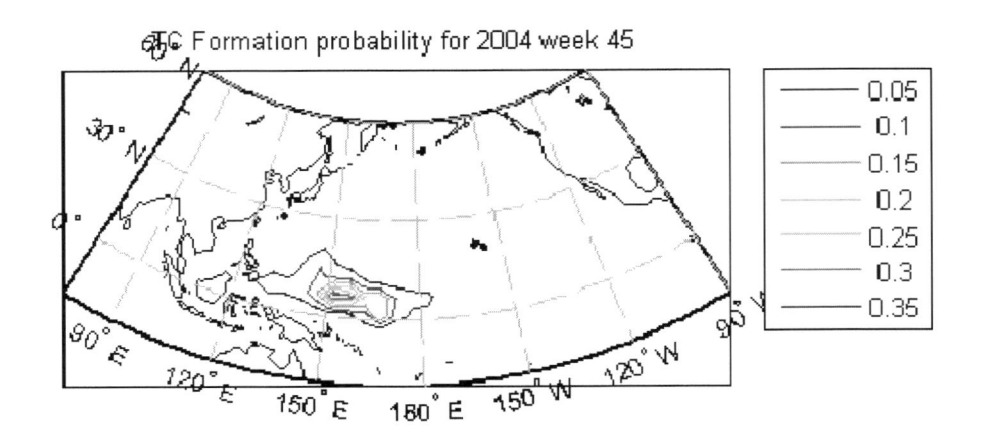

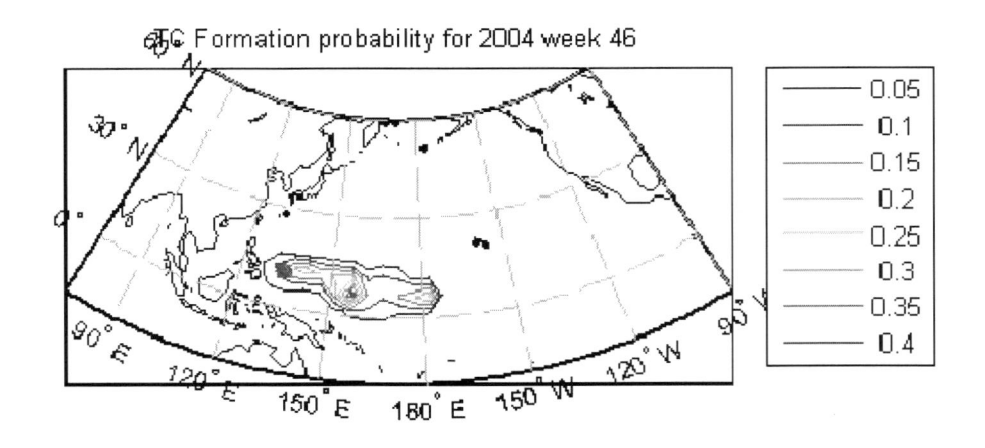

TC Formation probability for 2004 week 45

TC Formation probability for 2004 week 46

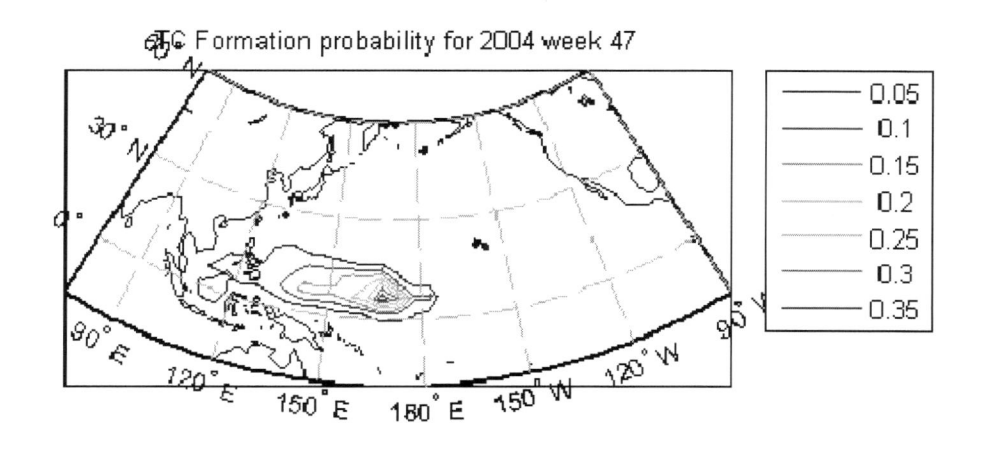

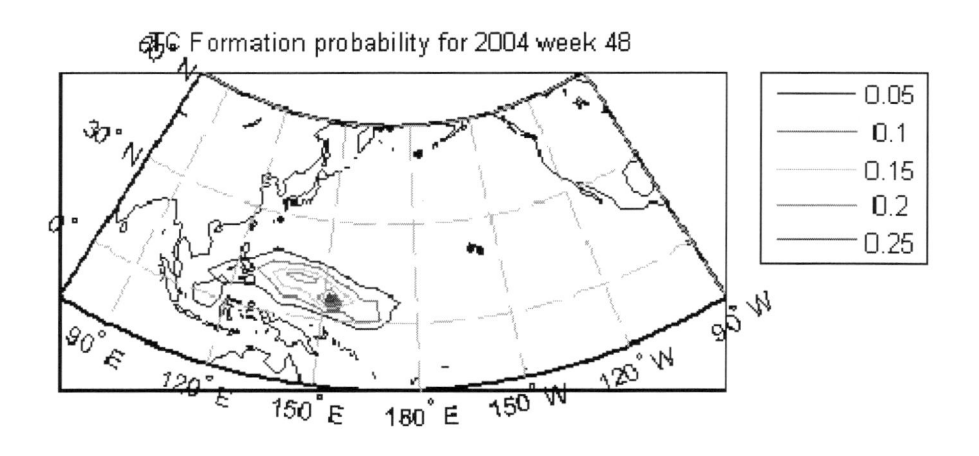

TC Formation probability for 2004 week 47

TC Formation probability for 2004 week 48

TC Formation probability for 2004 week 49

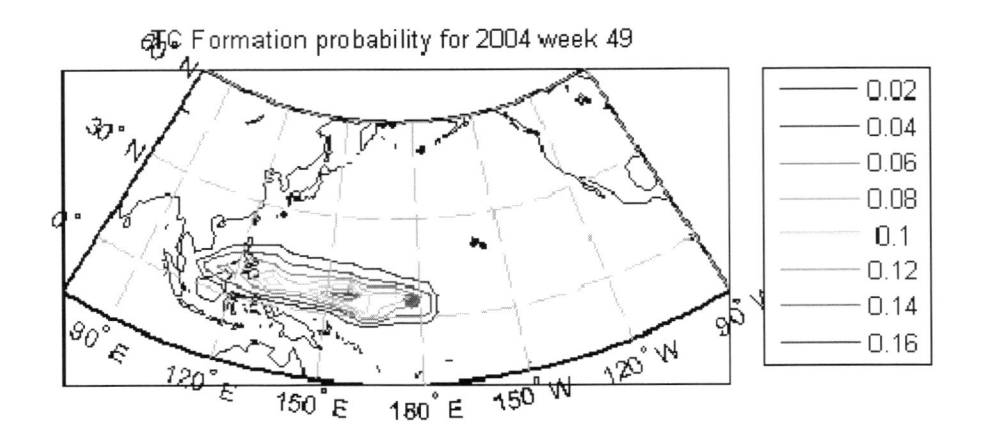

TC Formation probability for 2004 week 50

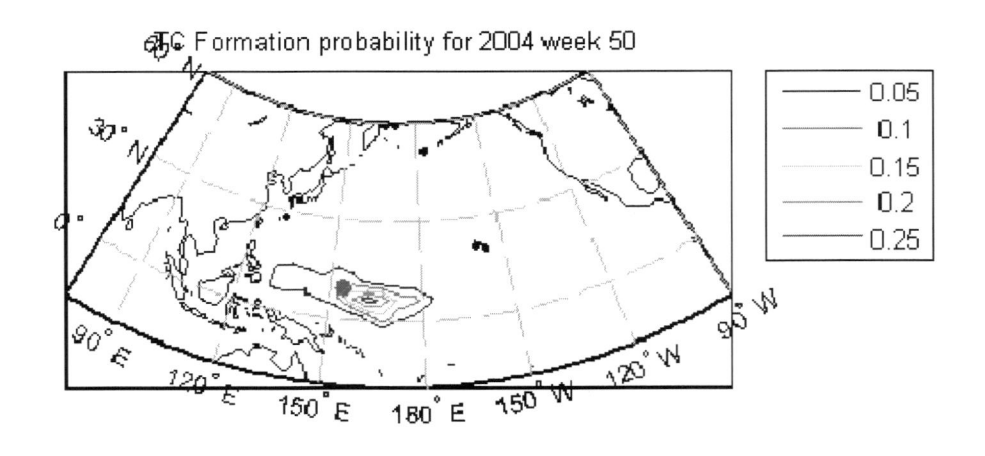

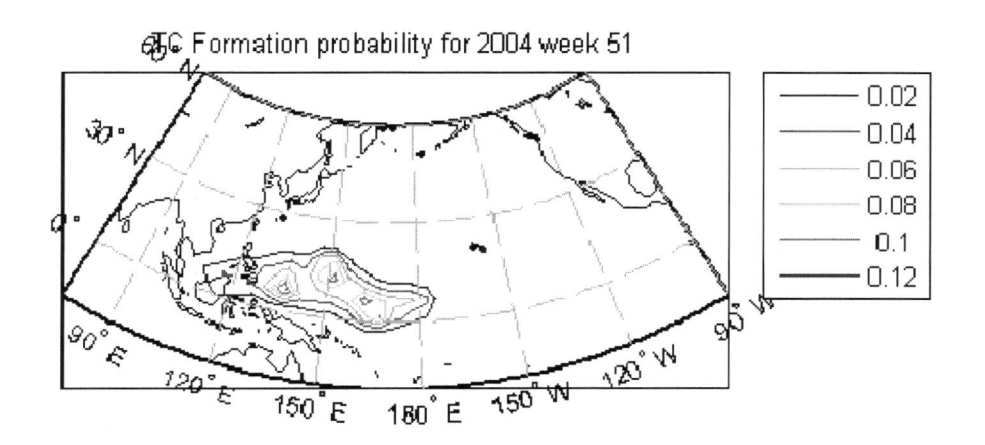

TC Formation probability for 2004 week 51

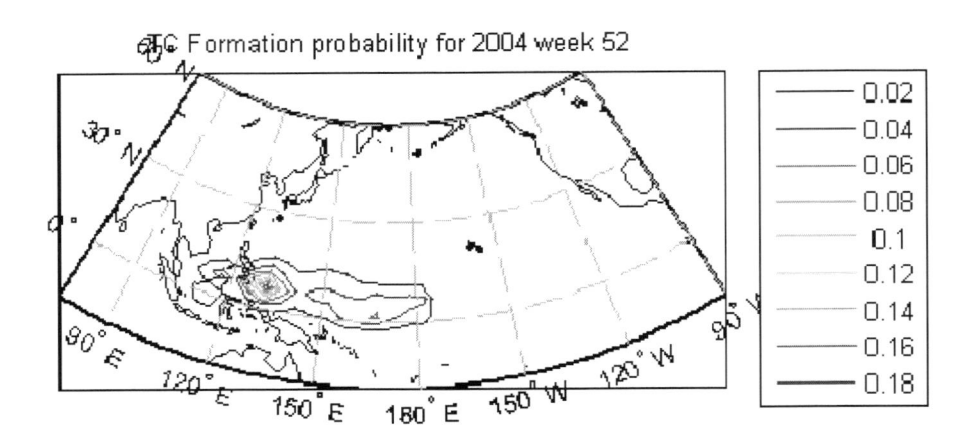

TC Formation probability for 2004 week 52

APPENDIX B. TC FORMATION PROBABILITY CONTOUR PLOTS FOR WEEKS 20-52 OF 1979

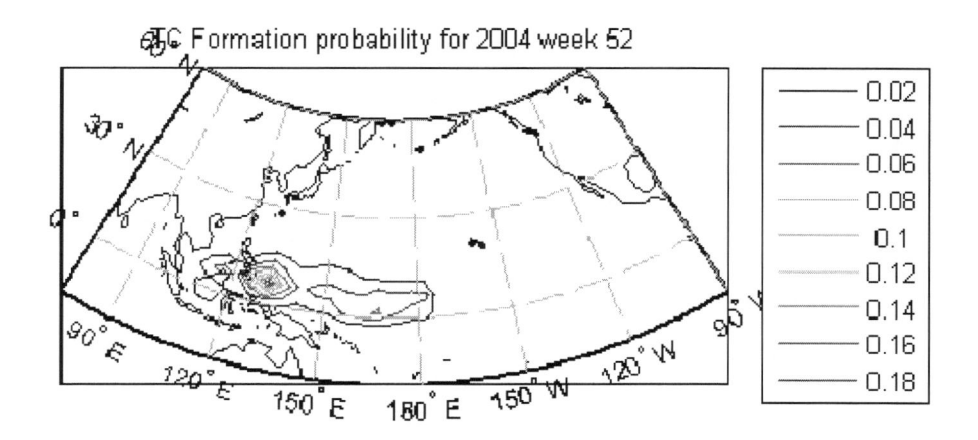

TC Formation probability for 2004 week 52

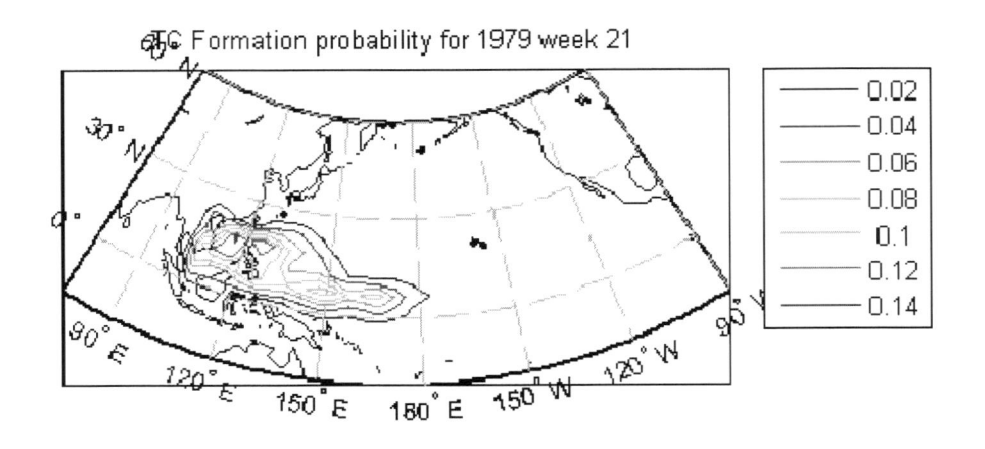

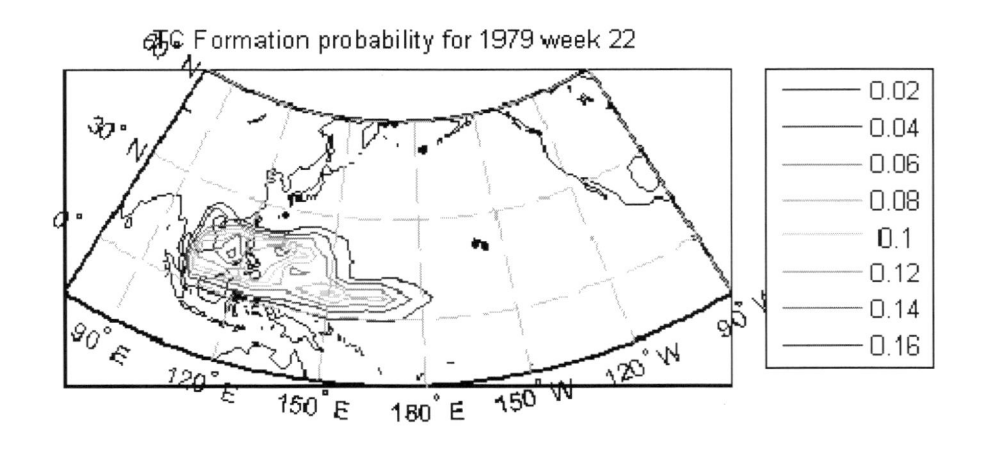

TC Formation probability for 1979 week 21

TC Formation probability for 1979 week 22

TC Formation probability for 1979 week 23

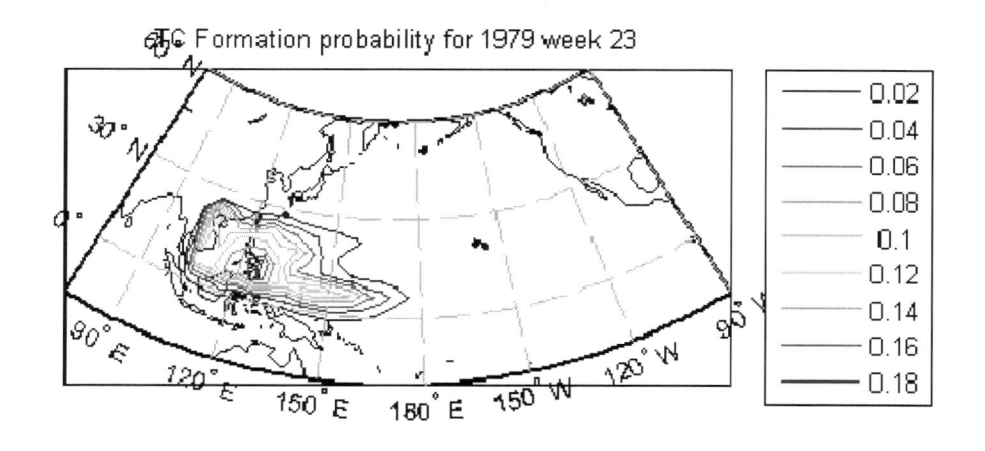

TC Formation probability for 1979 week 24

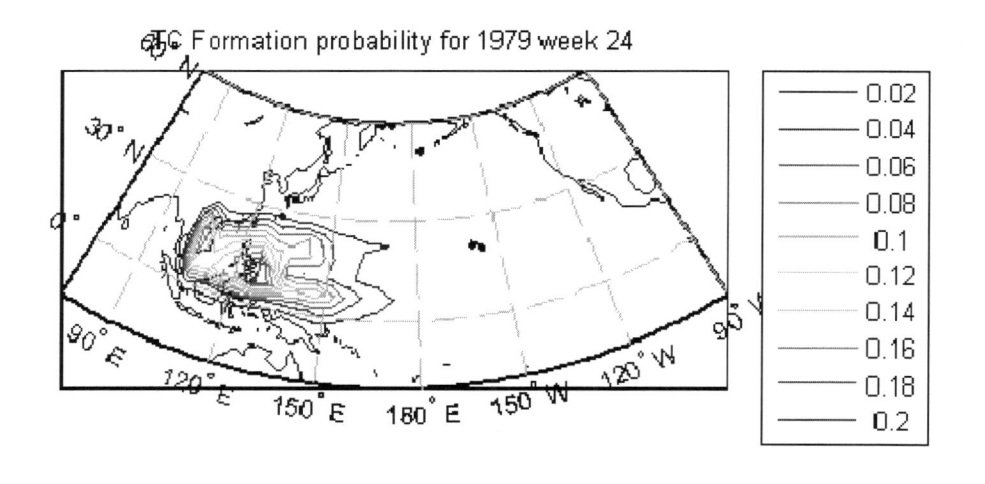

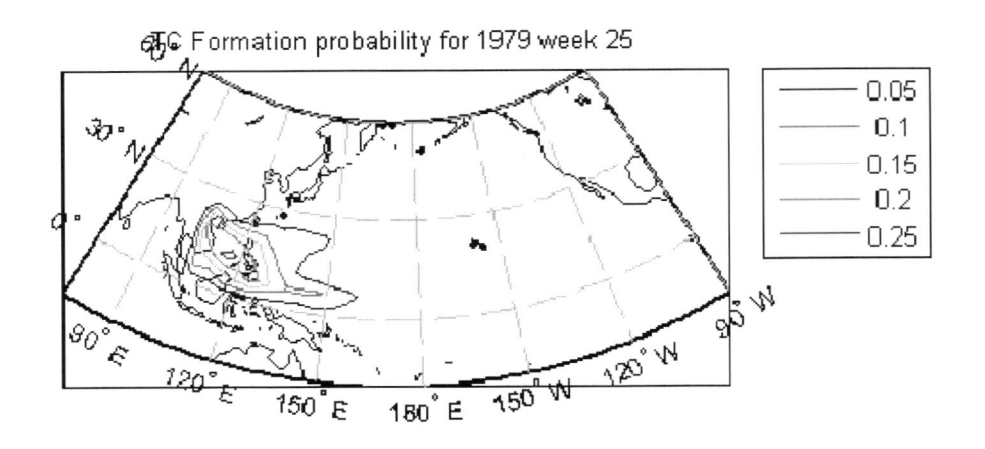

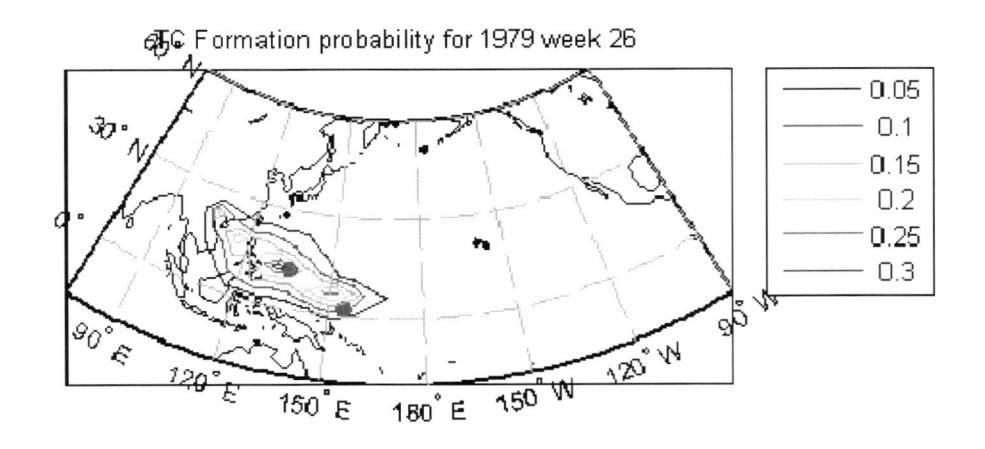

TC Formation probability for 1979 week 25

TC Formation probability for 1979 week 26

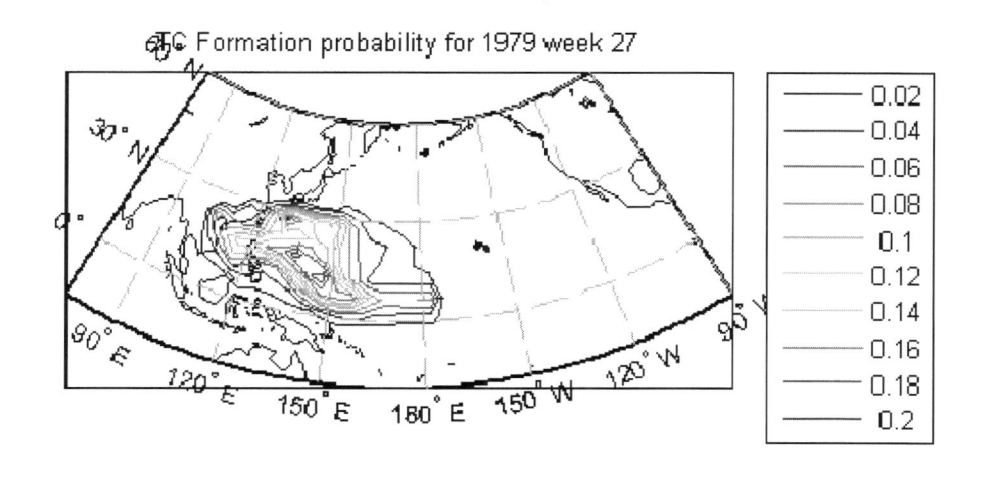

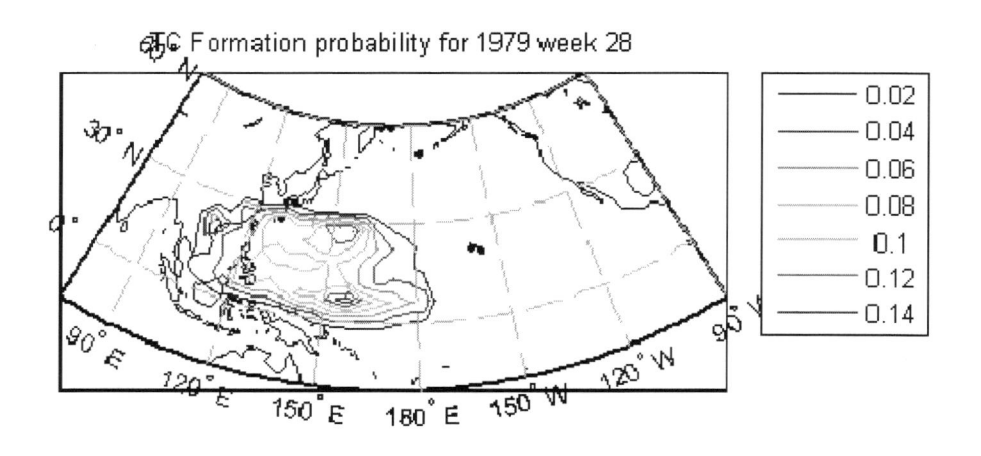

TC Formation probability for 1979 week 27

TC Formation probability for 1979 week 28

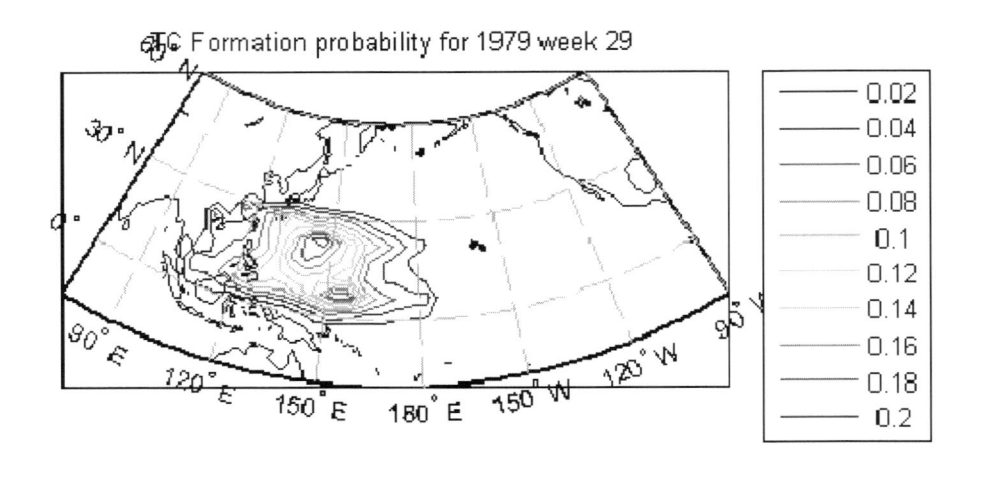

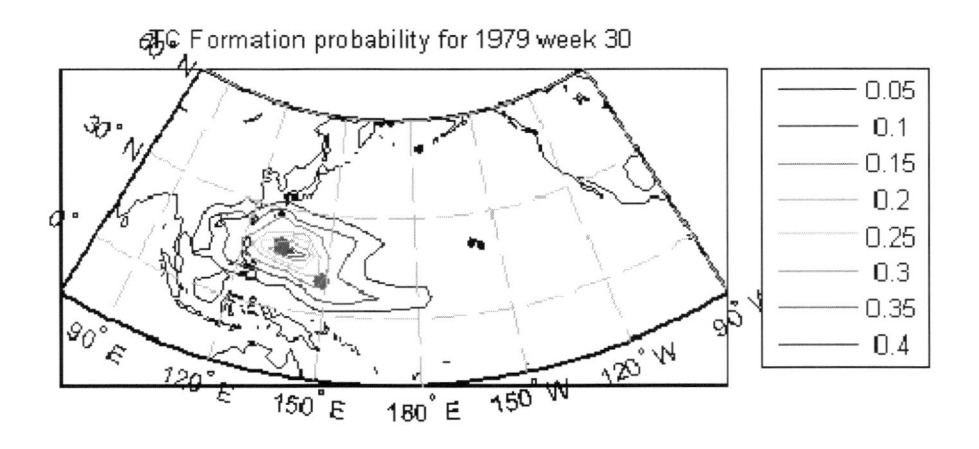

TC Formation probability for 1979 week 29

TC Formation probability for 1979 week 30

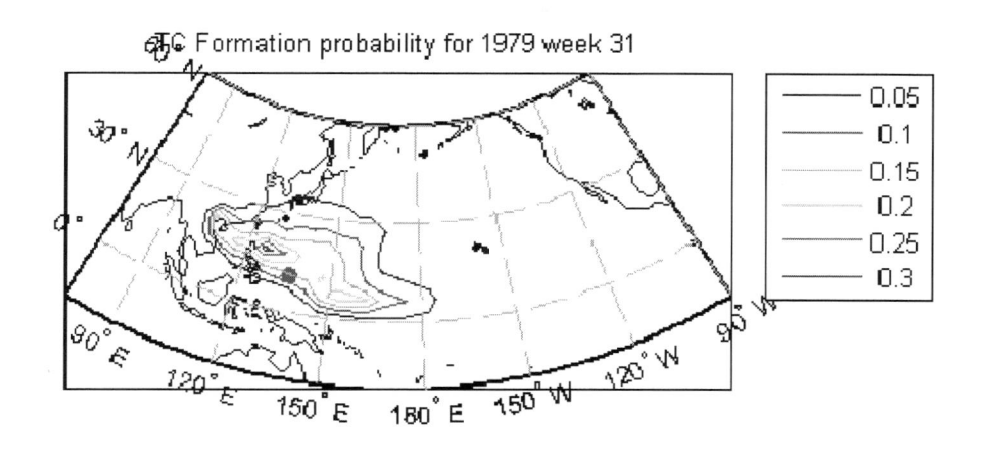

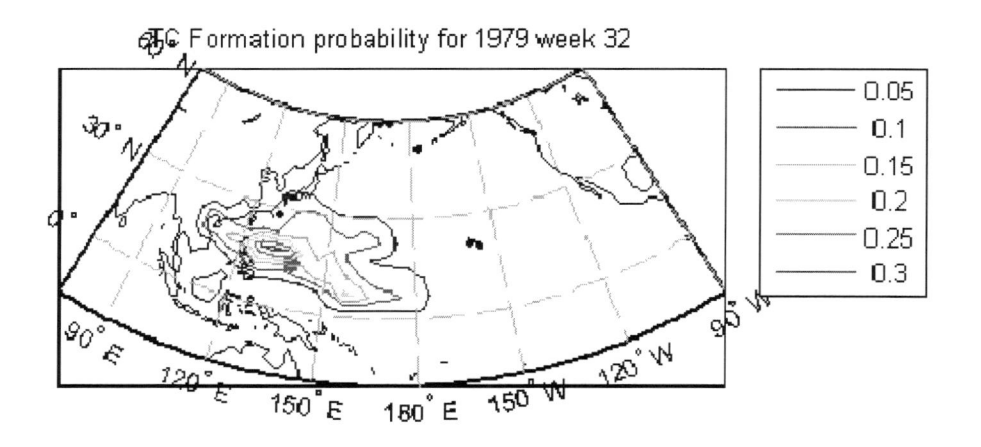

TC Formation probability for 1979 week 31

TC Formation probability for 1979 week 32

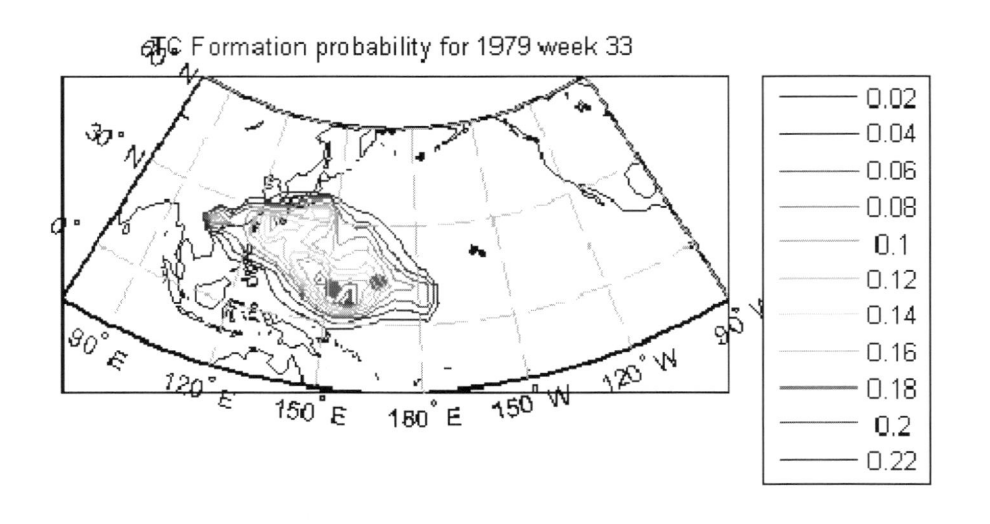

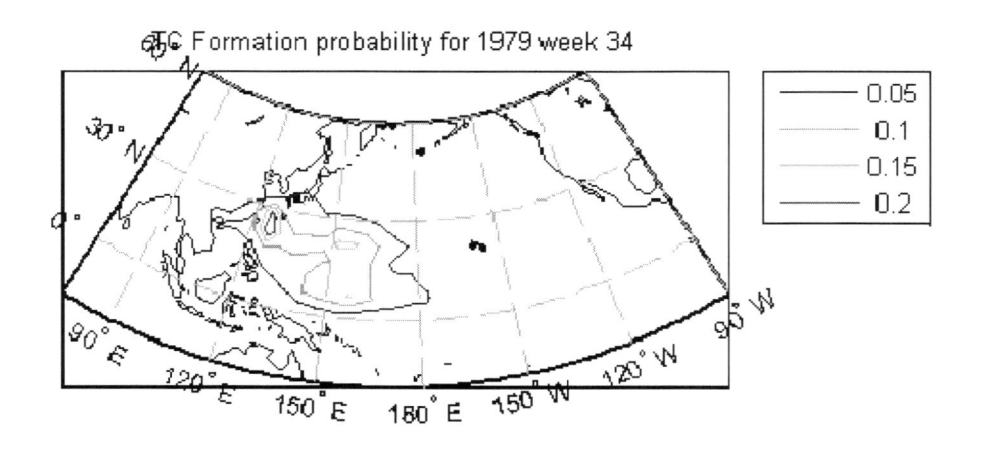

TC Formation probability for 1979 week 33

TC Formation probability for 1979 week 34

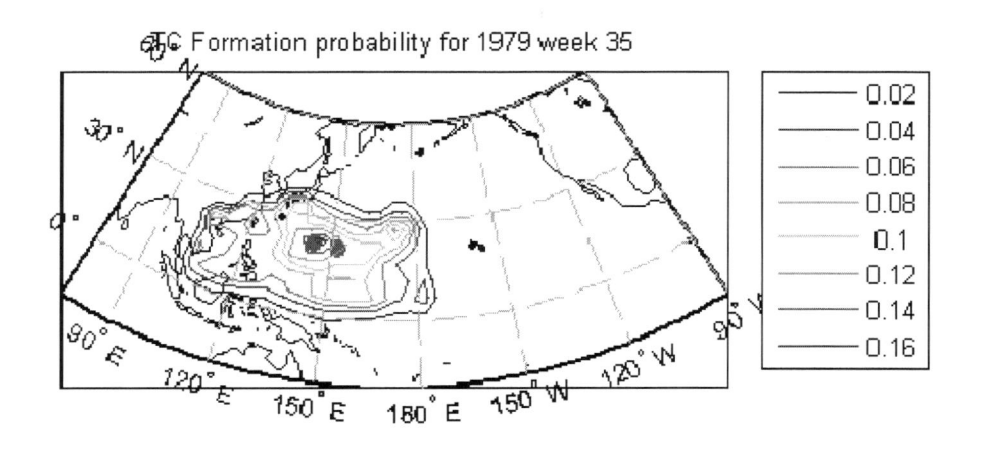

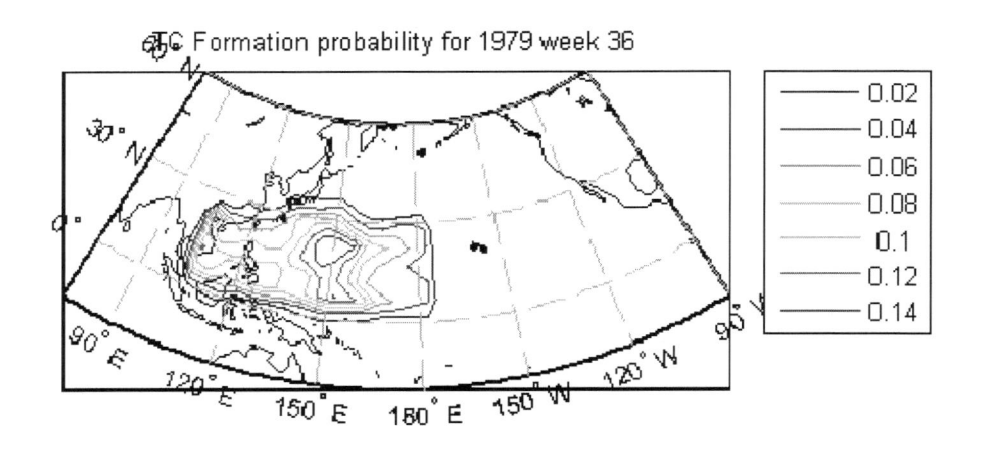

TC Formation probability for 1979 week 35

TC Formation probability for 1979 week 36

TC Formation probability for 1979 week 37

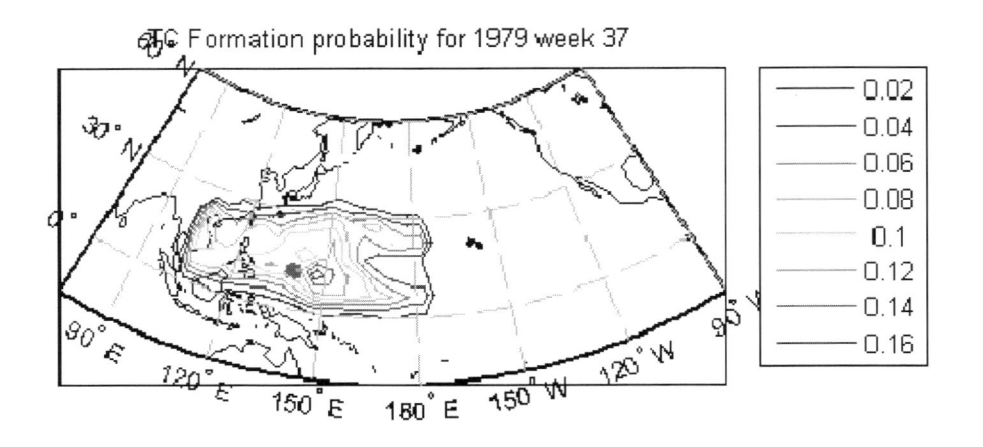

TC Formation probability for 1979 week 38

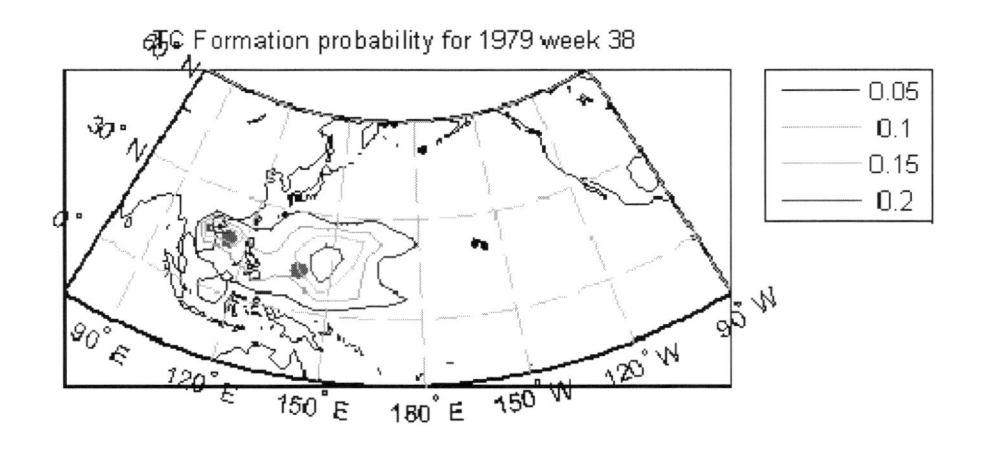

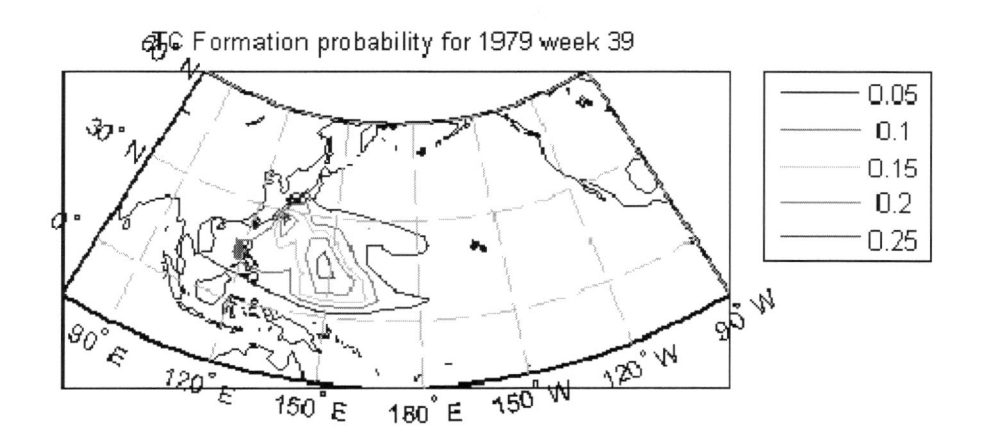

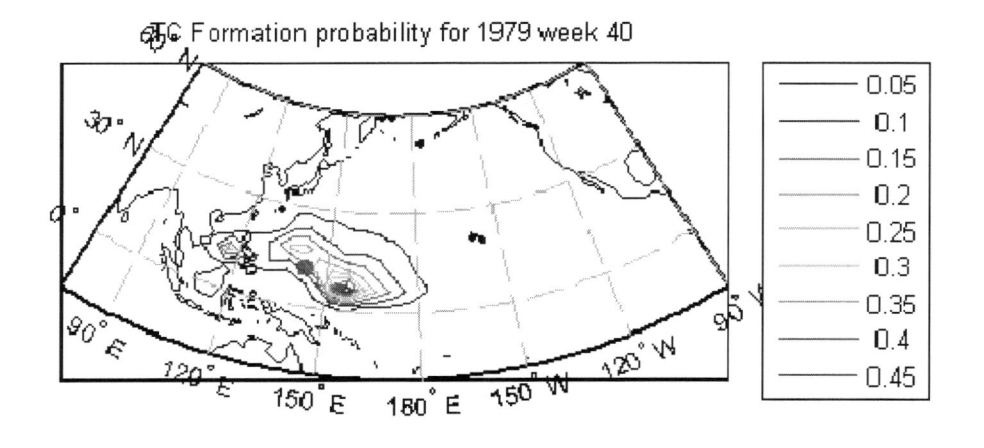

TC Formation probability for 1979 week 39

TC Formation probability for 1979 week 40

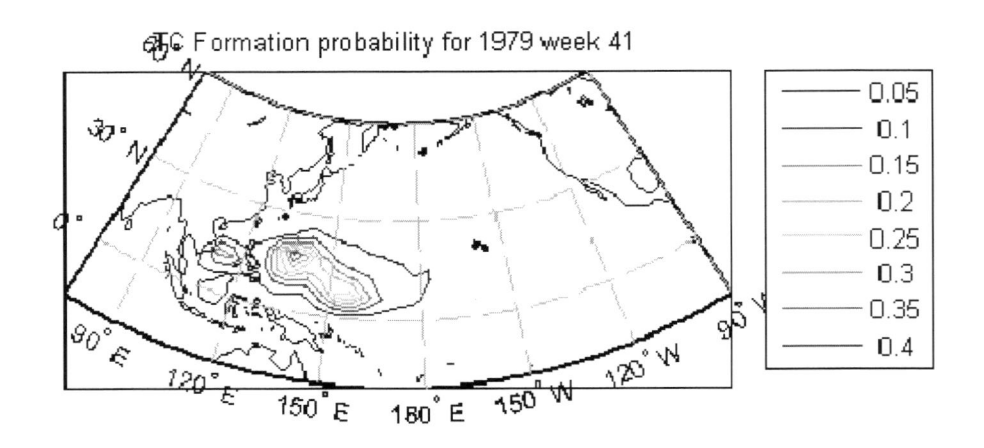

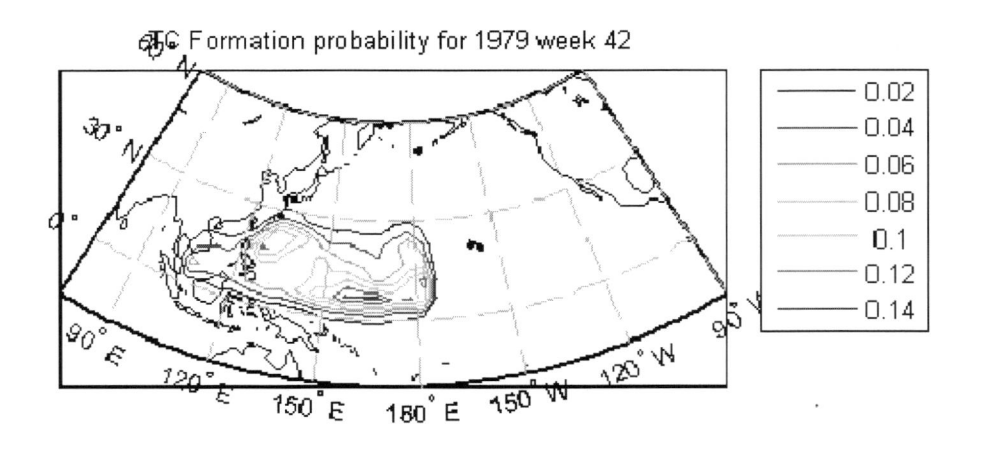

TC Formation probability for 1979 week 41

TC Formation probability for 1979 week 42

TC Formation probability for 1979 week 43

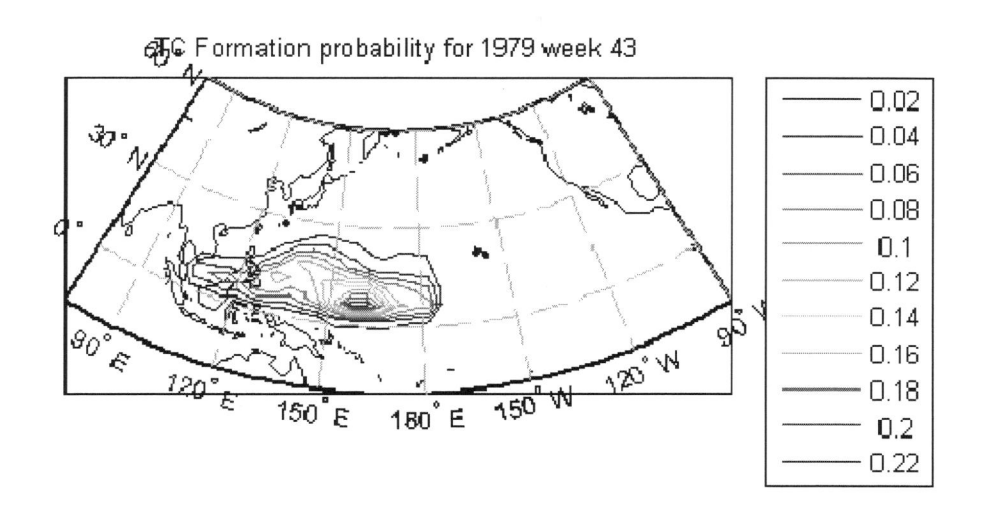

TC Formation probability for 1979 week 44

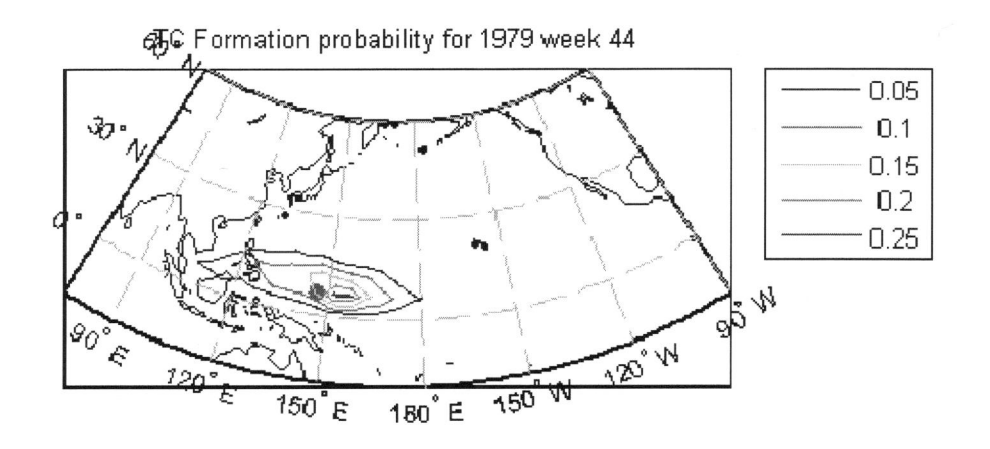

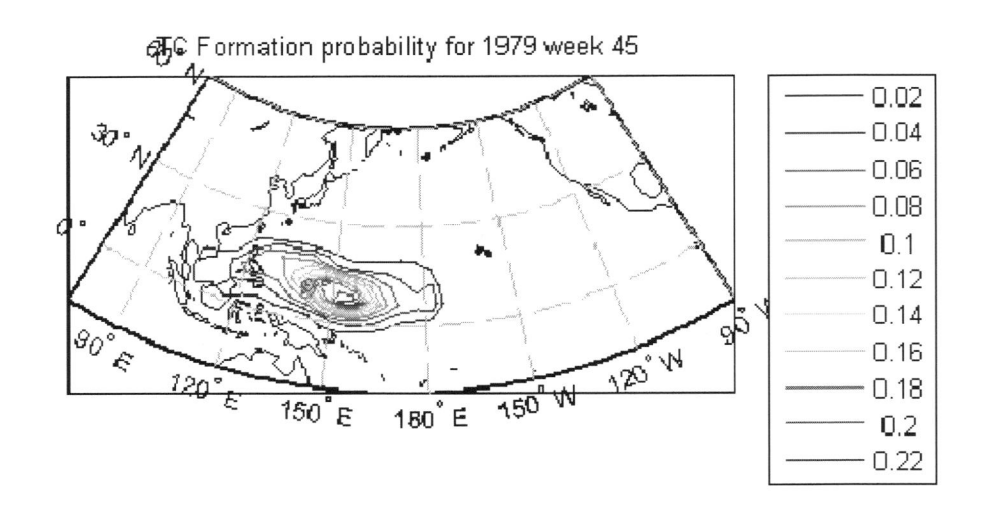

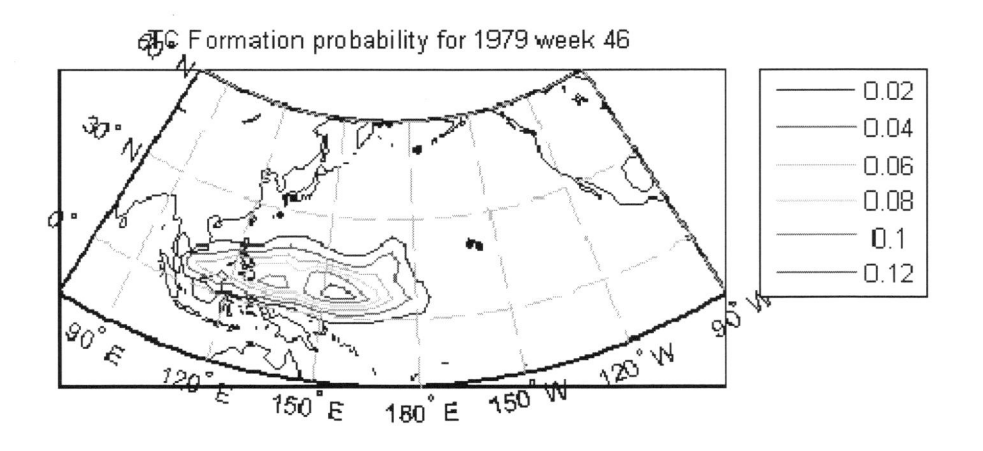

TC Formation probability for 1979 week 45

TC Formation probability for 1979 week 46

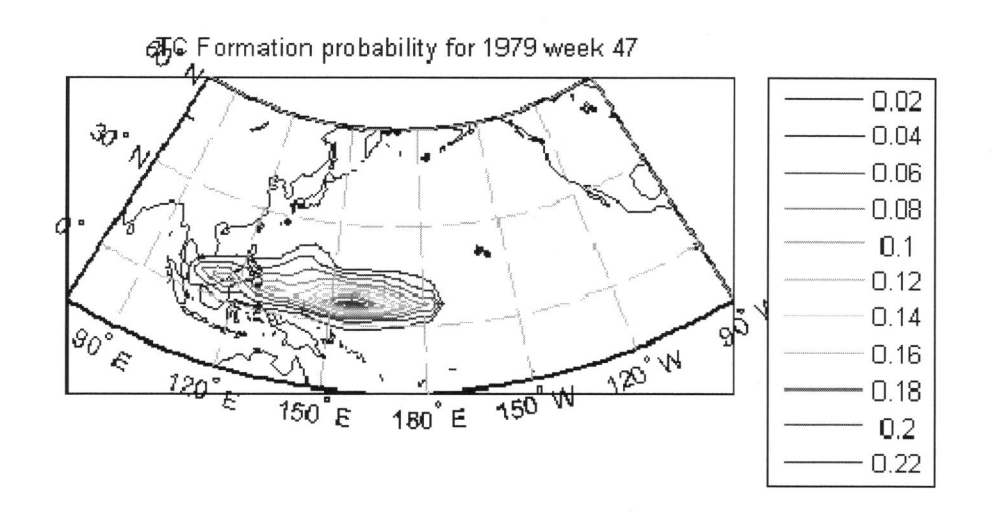

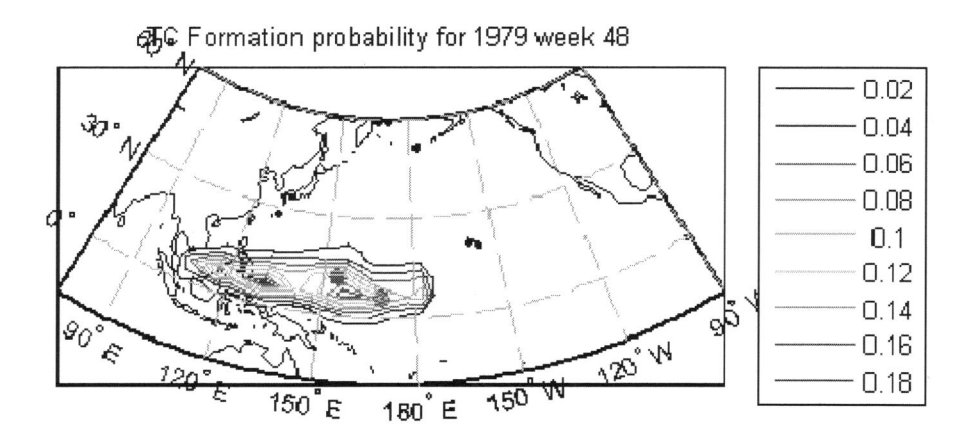

TC Formation probability for 1979 week 47

TC Formation probability for 1979 week 48

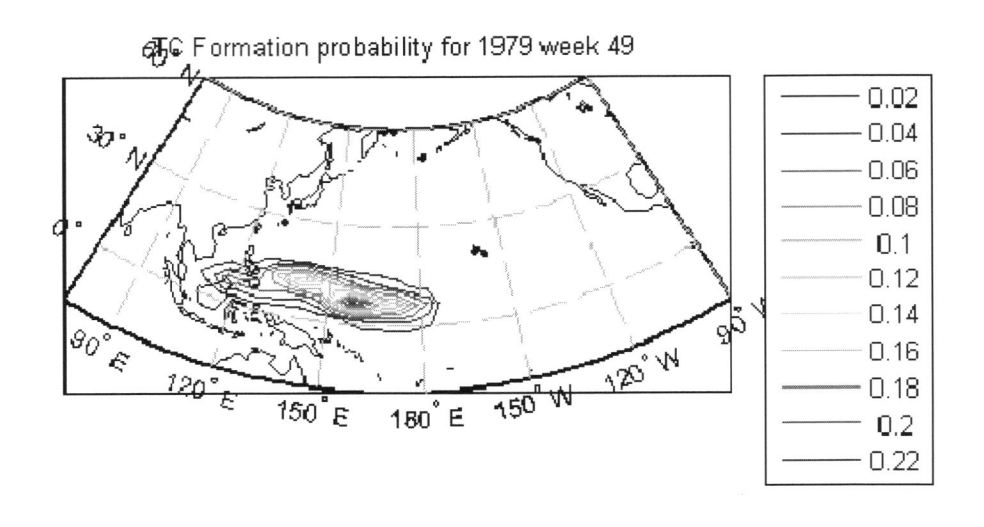

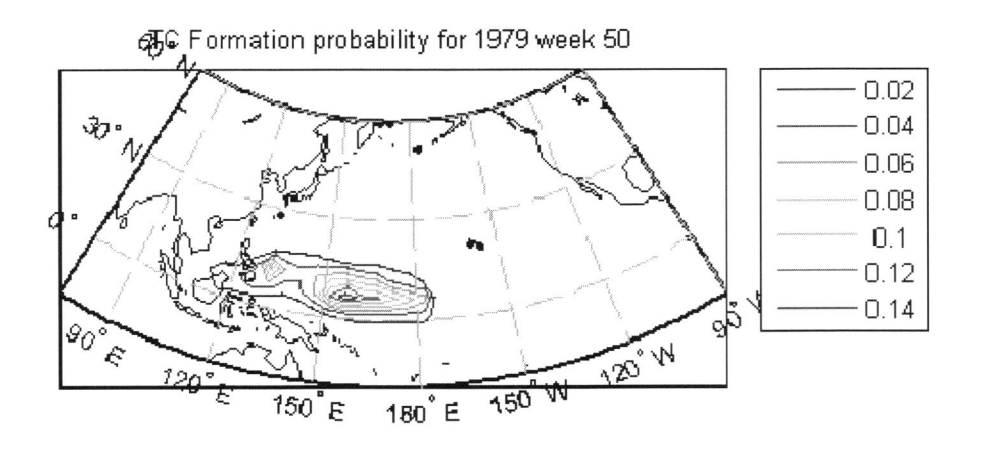

TC Formation probability for 1979 week 51

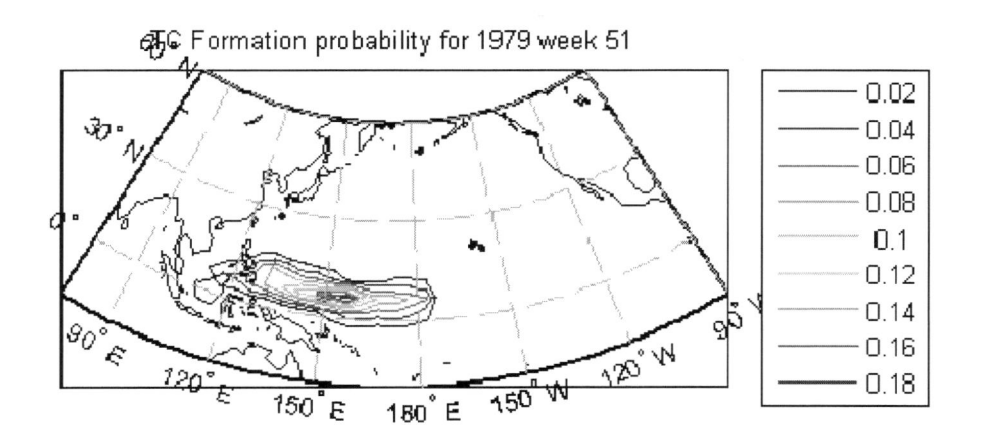

TC Formation probability for 1979 week 52

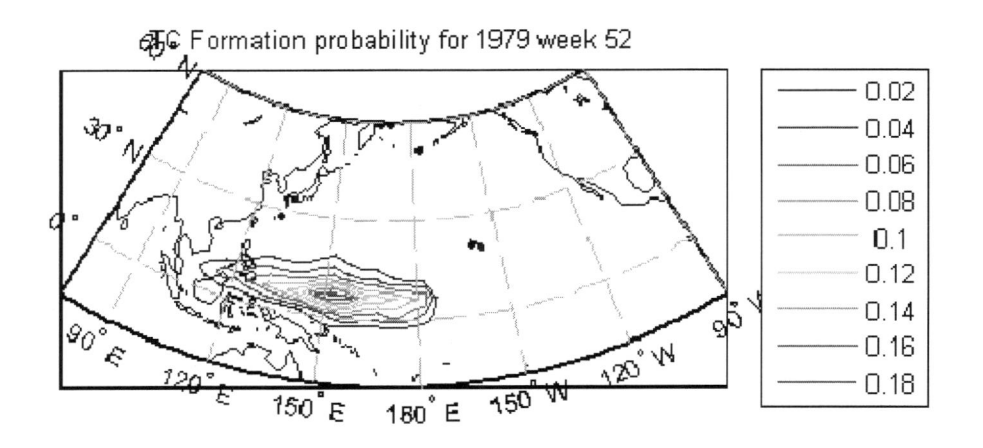

THIS PAGE INTENTIONALLY LEFT BLANK

LIST OF REFERENCES

Anthes, R. A., Corell, R. W., Holland, G., Hurrell, J. W., MacCracken, M. C., Trenberth, K. E., "Hurricanes and Global Warming – Potential Linkages and Consequences", *Bull.Amer.Meteor.Soc* **87**, 623 (2006).

Camargo, S. J., Robertson, A. W., Gaffney, S. J., Smyth, P., "Cluster Analysis of Western North Pacific Tropical Cyclone Tracks", downloaded 10 June 2007, available at www.datalab.uci.edu/papers/camargo_etal.pdf.

Chan, Johnny C. L., "Comment on "Changes in Tropical Cyclone Number, Duration, and Intensity in a Warming Environment"", *Science,* **311** (2006).

Chan, J. C. L., Liu, K. S., "Global Warming and Western North Pacific Typhoon Activity from an Observational Perspective", *Journal of Climate*, **17**, 4590-4602, (2004).

Chan, Johnny C. L., Shi, Jiu-en, Lam, Cheuk-man, "Seasonal Forecasting of Tropical Cyclone Activity over the Western North Pacific and the South China Sea", downloaded 21 June 2006, available at http://ams.allenpress.com/perlserv/?request=get-abstract&issn=1520-0434&volume=013&issue=04&page=0997.

Collett, D., 1991, *Modelling Binary Data.* Chap. 5, *Overdispersion.* Chapman and Hall, 2-6 Boundary Row, London SE1 8HN, UK, 188.

Curry, J. A., Webster, P. J., Holland, G. J., "Mixing Politics and Science in Testing the Hypothesis That Greenhouse Warming Is Causing a Global Increase in Hurricane Intensity", *Bull.Amer.Meteor.Soc* **87**, 1025-1037 (2006).

Devore, J. L., 2003: *Probability and Statistics for Engineering and the Sciences 6th ed.*, Brooks/Cole-Thomson Learning, 10 David Dr., Belmont, CA 94002.

Elsberry, Russell L., Frank, William M., Holland, Greg J., Jarrell, Jerry D., Southern, Robert L., 1971: "A Global View of Tropical Cyclones", Office of Naval Research, Arlington, VA 22217.

Emanuel, K., "Increasing Destructiveness of Tropical Cyclones Over the Past 30 Years", *Nature,* **436**, 686-688 (2005).

Emanuel, Kerry, "Anthropogenic Effects on Tropical Cyclone Activity", downloaded 21 June 2006, available at http://wind.mit.edu/~emanuel/anthro2.htm.

Frank, William M., Young, George S., "The 80 Cyclones Myth", The Pennsylvania State University, downloaded 15 August 2006, available online at: http://ams.confex.com/ams/pdfpapers/108554.pdf.

Gray, W. M., 1975: "Tropical Cyclone Genesis in the Western North Pacific", Naval Air Systems Command, Washington DC, 20361.

Intergovernmental Panel on Climate Change, Working Group I, the Physical Basis For Climate Change, downloaded 15 June 2007, available online at: http://ipcc-wg1.ucar.edu/wg1/wg1-report.html.

Jury, Mark R., Pathack, Beenay, Parker, Bhawoodien, "Climatic Determinants and Statistical Prediction of Tropical Cyclone days in the Southwest Indian Ocean", *Journal of Climate*, **12**, 1738-1746 (1999).

Kalnay et al., "The NCEP/NCAR 40-year reanalysis project", *Bull. Amer. Meteor. Soc.*, 77, 437-470, 1996.

Klotzbach, P. J., Gray, W. M., 2006, "Extended Range Forecast of Atlantic Seasonal Hurricane activity and U.S. Landfall Strike Probability for 2006", Department of Atmospheric Science, Colorado State University, Fort Collins, CO, 80523.

Klotzbach, Philip J., "Trends in global tropical cyclone activity over the past twenty years (1986-2005)", *Geophysical Research Letters,* **33**, L10805 (2006).

Landsea, C.W., "Counting Atlantic Tropical Cyclones Back to 1900", *Eos,* **88**, 197-208 (2007).

McBride, J. L., "Global Perspectives on Tropical Cyclones", 1995: *World Meteorological Organization* **TCP-38**, Ch. 3, Secretariat of the World Meteorological Organization - Geneva, Switzerland, 1995.

Mann, M. E., Emanuel, K. A., "Atlantic Hurricane Trends Linked to Climate Change", *Eos*, **87**, 233, 238, 241 (2006).

Palmen, E., Newton, C. W., 1969: *Atmospheric Circulation Systems, Their Structure and Physical Interpretation.*

Pielke, R. A. Jr., Landsea, C., Mayfield, M., Laver, J., Pasch, R., "Hurricanes and Global Warming-Potential Linkages and Consequences", *Bull.Amer.Meteor.Soc* **86**, 1571-1575 (2005).

Pielke, R. A. Jr., Landsea, C., Mayfield, M., Laver, J., Pasch, R., "Reply to "Hurricanes and Global Warming-Potential Linkages and Consequences"", *Bull.Amer.Meteor.Soc* **87**, 628-631 (2006).

Powell, Mark D., Reinhold, Timothy A., "Tropical Cyclone Destructive Potential By Integrated Kinetic Energy", *Bull.Amer.Meteor.Soc* **88**, 513-526 (2007).

Rayner, N. A.; Parker, D. E.; Horton, E. B.; Folland, C. K.; Alexander, L. V.; Rowell, D. P.; Kent, E. C.; Kaplan, A. "Global analyses of sea surface temperature, sea ice, and night marine air temperature since the late nineteenth century" *J. Geophys. Res*.Vol. 108, No. D14, 4407 10.1029/2002JD002670.

Trenberth, K. E., "Uncertainty in Hurricanes and Global Warming", *Science*, **308**, 1753-1754, (2005).

Trenberth, K. E., Shea, D. J., "Atlantic hurricanes and natural variability in 2005", *Geophysical Research Letters*, **33**, (2006).

Webster, P. J., Holland, G. J., Curry, J. C., Chang, H.-R., "Changes in Tropical Cyclone Number, Duration, and Intensity in a Warming Environment", *Science,* **309**, 1844 (2005).

Webster, P. J., Holland, G. J., Curry, J. C., Chang, H.-R., "Response to Comment on "Changes in Tropical Cyclone Number, Duration, and Intensity in a Warming Environment"", *Science,* **311**, 1713 (2006).

Wu, M.C., Yeung, K.H., Chang, W.L., "Trends in Western North Pacific Tropical Cyclone Activity", *Eos,* **87**, 537-538 (2006).

Printed in Great Britain
by Amazon.co.uk, Ltd.,
Marston Gate.